GENESIS

THE ORIGINS OF MAN AND THE UNIVERSE

ALSO BY JOHN GRIBBIN
FROM DELACORTE PRESS/
ELEANOR FRIEDE

WHITE HOLES:
COSMIC GUSHERS IN
THE UNIVERSE

TIMEWARPS

THE DEATH OF THE SUN

GENESIS

THE ORIGINS OF MAN AND THE UNIVERSE

JOHN GRIBBIN

DELACORTE PRESS / ELEANOR FRIEDE

Published by
Delacorte Press/Eleanor Friede
1 Dag Hammarskjold Plaza
New York, N.Y. 10017

"The Age of the Universe" and "Growth and Limits"
(as "What Future for Futures?") first appeared
in *New Scientist*, London, the weekly review
of science and technology.

Manufactured in the United States of America

First printing

Designed by Jo Anne Bonnell

Original drawings by Neil Hyslop

Library of Congress Cataloging in Publication Data

Gribbin, John R
    Genesis : the origins of man and the universe.

    Bibliography: p.
    Includes index.
    1. Astronomy.   2. Life—Origin.   I. Title.
QB44.2.G753   577   80-24267
ISBN 0-440-02832-9

For Jo and Ben
who may one day be involved
in developing the *next* world picture

ACKNOWLEDGMENTS

This is the book I always wanted to write, even before I knew I wanted to write books, and most of my life can be regarded as an apprenticeship leading up to the production of the volume you hold in your hands. In a sense, then, everyone I have ever asked a question of (which means everyone I have ever met) should be thanked for their contributions to my store of information. My parents, who could have stifled the early flicker of curiosity if they had been less helpful and understanding, deserve special mention, along with the books of George Gamow and the TV appearances of Jacob Bronowski, which made a major impact at an impressionable age (yes, we did have TV that long ago). Rather later in life, I was saved from a floundering uncertainty about what to do with a degree in physics by the discovery that it was actually possible to go on to specialization in cosmology—the study of the Universe—and by the presence of Bill McCrea at Sussex University, where as first head of the University's Astronomy Center he inspired my continuing fascination with "the beginning." At Cambridge University, John Faulkner was the first academic I worked with who shared my view that science should be fun, a belief which is, unfortunately, rare enough in academic circles to be worthy of mention. Over a period of more than ten years I have kept more or less up-to-date on matters cosmological, in spite of moving on to other things, in large measure through the willingness of Martin Rees to discuss the latest developments or, in recent years, to divert my questioning on to his colleagues at the Institute of Astronomy in Cambridge, where he is now such a big wheel that they have no choice but to comply with his "suggestions" that they talk to me (which, of course, they always do happily and most helpfully; thanks in particular to David Hanes, Simon White, and Michael Fall).

Although I became interested in the modern theory of plate tectonics largely through sharing a computer in Cambridge with the geophysics group, it wasn't until I was working for the science journal *Nature* in the early 1970s that I discovered this was a revolutionary new theory; my geophysical education has therefore been somewhat backward in that I learned the latest ideas first, but this may have been no hindrance! Dan McKenzie in Cambridge and Peter Smith, of the Open University, have at various times made major contributions to this backward education. It was at *Nature*, too, that I had the opportunity to discover new developments in the understanding of climatic change, and managed over the years to pick the brains of Hubert Lamb and Michael Kelly, of the University of East Anglia, and Stephen Schneider and colleagues at the U.S. National Center for Atmospheric Research, to extend my knowledge of the "Earth sciences" upward from the ground into the atmosphere.

All that, though, only accounts for roughly half of this book, and the most exciting development in my continuing education in recent years has been the opportunity, at last, to find out something in detail about the nature, origin, and evolution of life on Earth. David Attenborough's superb TV series helped to set me off down this particular trail, reviving an older interest. But such a presentation inevitably raises as many questions as it answers, and I would have been left floundering in my efforts to get deeper into the subject without the help of Roger Lewin (now with *Science*) and Jeremy Cherfas (now at the University of Oxford), of *New Sci-*

entist. The prospect of having found an area of knowledge new to me and as big, as the balance of this book shows, as everything else I have been involved with put together makes theirs the major recent contribution to my continuing program of self-education, with the promise (threat?) of much more material to be mined from this vein.

But the single most important contributor to this particular book has been my wife, a researcher of rare skill who both physically obtains the material input (in the form of books, articles, and so on) for my writing process and tells me, without fear or favor, when the output is incomprehensible. Without her, the book wouldn't exist at all.

Thanks, then, to Lilla, Bill, George, Jacob, Bill, John, Martin, Dan, Peter, Hubert, Mick, Steve, David, Roger, Jeremy, and especially Min; but the mistakes, as they say, are entirely my own work.

The most incomprehensible thing
about the Universe is that it is comprehensible.

—Albert Einstein

CONTENTS

GENESIS

THE ORIGINS OF MAN AND THE UNIVERSE

INTRODUCTION

The question "where do we come from?" is the most profound question it is possible to ask, and the ability to ask that question is as good a criterion as any used to distinguish between intelligent and nonintelligent species. This curiosity extends to our immediate surroundings, since human origins cannot be considered in isolation, but only as part of a greater mystery encompassing the origins of life on Earth and the place of the Earth in the Universe of stars and galaxies we see around us. A bear, say, may have a natural interest in the fact that bees make honey and nest in trees, but as far as we know the bear does not ponder on the mysteries of why bees nest in trees, nor does he speculate as to the origin of the hexagonally shaped cells with which their honeycombs are made. It is a distinguishing feature of human life alone, as far as life on Earth is concerned, to ask the where and how of our origins and surroundings, as well (with rather less success) as *why* the Universe should be the way it is.

In this book I have attempted to provide, for a reader with no background in science, an overview of the best modern, scientific answer to the question "where do we come from?" Because the story of life on Earth cannot be divorced either from the story of the Earth as a planet or from the story of our Sun, on which we depend for light and heat, as a star, the overview has to range across many scientific disciplines, as well as across vast reaches of space and time. And since the story of the Sun and the Earth depends on the nature of the Galaxy of stars around us (the Milky Way) and of the Universe itself, it seems to me appropriate to take the story right back to the beginning of the Universe itself, the Big Bang of creation which occurred some 15,000 million years ago.

This is possible only because the past two decades have seen revolutionary changes in our understanding of the Earth and the Universe at large (not least being the discovery that there really *was* a definite beginning to everything!), and dramatic developments in the study of human origins and evolution, with discoveries of fossil remains in East Africa and new developments in molecular biology combining to shift the establishment view of how we got to be the way we are more significantly than at any time since Darwin presented his ideas on evolution. The result of all these changes is that a new picture has emerged both of mankind and of the Universe in which we live. But all too often, alas, the separate parts of this emerging picture have remained separate. Science has become so specialized and compartmented, with such a mass of detail for any specialist to absorb in his own specialty, that an anthropologist, say, may know nothing of astronomy beyond what he learned in school (which may now be well out-of-date!) while a cosmologist knows no more about the new picture of human evolution than he gleans from the latest TV special.

Of course there are exceptions, and it is possible to find an astronomer, Carl Sagan, writing about the evolution of human intelligence. But what the specialization tends to conceal is the way that all of the threads can be woven into a new overall picture of mankind, the Earth, and our place in the Universe. The completeness of the resulting picture is surely as significant as the detail of any part of the whole, suggesting that we really do have a good understanding of what is going on in the Universe and how we got to be the way we are. But one word of warning is in order. The picture is the best we have, and seems more complete than ever before. It is unlikely, though, to be the final picture of its kind. At the end of the nineteenth century, many scientists felt that all the great discoveries had been made, with their remaining work being simply to dot the *i*'s and cross the *t*'s of science. They understood the nature of the chemical elements, the "billiard ball" model of atoms seemed satisfactory, and so on. Yet in the early twentieth century this cozy picture was turned on its head by relativity theory, the ideas of quantum mechanics, and later by the realization that the supposedly "elementary" atomic particles disguised the existence of complex hierarchies of other elementary particles.

It is just possible that our present, seemingly complete, picture of the Universe indicates not a final understanding but the end of another phase of scientific development, comparable to the picture of late nineteenth-century science. However, it is the best picture we have ever had of the way things are, and if in ten years from now the applecart is overturned again, it might still be of interest to know how things stood at the beginning of the 1980s. At least my story might then provide amusement for the next generation, looking back on our quaint, old-fashioned ideas! Having decided to attempt the task of explaining the origin of "life, the Universe, and everything"* in one volume, the main problem is where to begin. Knowledge was built up from nearby, familiar objects outward, and many astronomy books still follow the traditional pattern, working outward from our home here on Earth to the Sun on which we depend, then continuing outward on a cosmic scale to other planets in the Solar System, other stars in our Galaxy, and the Universe as a whole. This is a logical but Earth-centered approach which encourages, without explicitly saying so, the idea that we are somehow special and occupy a special place in the Universe. And it also suffers from another curious distortion, which is a result of the way we see the Universe.

Virtually all of our information about the Universe at large

---

* A cult BBC radio show of the late 1970s, "The Hitchhiker's Guide to the Galaxy," revealed that the answer to "life, the Universe, and everything" was forty-two, but that nobody knew the *question*. Sir Fred Hoyle has written: "The answers are not important, the questions are." My version of the story of life, the Universe, and everything may not be as succinct as the Hitchhiker's Guide, and may not fit Sir Fred's aphorism, but I have tackled it in the same spirit Mallory tackled Everest—"because it is there."

comes from analysis of electromagnetic radiation—light, radio waves, X-rays, or whatever—all of which travels at the speed of light, 3×10^{10} centimeters a second. Although this is a very large speed (3 followed by 10 zeroes cm in one second), the scale of the Universe is also very large, so that electromagnetic radiation takes a long time to reach us from other stars and galaxies. Even for a relatively nearby star, light spends years on the journey across space to Earth, so that we see the star as it was years ago, when the light left it. Farther away across the Universe, we can detect light from galaxies so remote that we see them as they were millions, hundreds of millions, or even thousands of millions of years ago, by light which has been that long on the journey. So, in a very real sense, as we probe farther away in space using our modern astronomical instruments, we are looking farther back in time.

In itself this is a boon to astronomers, who can test some of their ideas about how the Universe has evolved by comparing nearby regions with those farther away, in effect comparing a recent snapshot of the Universe with one from its youth. But it does mean that telling the story of the Universe by working outward from here and now on Earth is telling the story, in a sense, backward in time. Having had the chutzpah to tackle the task of telling the whole story of the origins of man and the Universe, it seemed only right, to me, to do the job properly, and to begin where all good stories begin—in the beginning.

John Gribbin
May 1980

1
THE ORIGIN OF
THE UNIVERSE

In the beginning, there was nothing at all. This is a very difficult concept, and one which causes a great deal of misunderstanding among many people who have heard of the idea of the "Big Bang," the creation of the Universe as we know it in some vast explosion of matter and energy. From our everyday experience, we know what a big bang is like—a concentration of matter, triggered by some energetic process, blasting outward into space. Even many astronomers, I suspect, have as their own personal image of the Big Bang the explosion of a star (a supernova) magnified by as great a degree as their imaginations will allow.

But before the Big Bang of creation, there wasn't even any empty space. Space and time, as well as matter and energy, were created in that "explosion," and there was no "outside" for the exploding Universe to explode into, since even when it was only just born and beginning its great expansion, the Universe contained everything, including all the empty space. Einstein's revolutionary new vision of the Universe, developed in the early part of this century, did two things. It unified matter and energy as two aspects of a greater whole, with the implication that matter can be converted into energy in the right circumstances, circumstances that we now know exist inside stars and nuclear bombs. And it also unified space and time as different facets of another greater whole, space-time, an underlying fabric which provides the basic structure of the Universe, within which the much less significant matter-energy is carried along for the ride, interacting with space-time through gravitational forces.

A common analogy to represent the expanding Universe (and just how we know it is expanding I will come to in a moment) is provided by the imaginary example of a rubber sheet, marked with ink blobs, which is slowly being stretched. The rubber sheet represents space-time, what we used to think of as empty space. The ink blobs represent concentrations of matter, galaxies of stars which move apart as the rubber "universe" expands.* And the important point of the analogy is that the "galaxies" in the rubber sheet universe move apart from one another, mimicking the way real galaxies behave in the real Universe, but they do not move through the fabric of the rubber, just as real galaxies do not move through the fabric of space-time.

All this, in a way, is getting ahead of the story. The key to the science of cosmology today is the discovery that the Universe is indeed expanding, and by discussing the nature of the expansion before presenting the evidence that it occurs I am clearly putting the cart before the horse. But it does fit my theme of starting at the beginning, by bringing out the point that the Big Bang saw the beginning of everything. Not just matter and energy, exploding out into the void, but the void itself—space. And not just space, but its counterpart, time, the other facet of the space-time fabric. The flow of time as we know it also began with the Big Bang so that it may be literally meaningless to ask what happened "before"

---

* A convenient convention used in astronomy and cosmology is to capitalize names which apply to our own Earth, Moon, Solar System, and the Universe in which we live, while lowercase is used for the moons of Mars, say, and for an imaginary universe (even one as simplistic as a rubber sheet!) which is used as a "model" to bring out some facet of a theorist's thinking. I will follow this convention; in particular, any description of the Universe means the best understanding of how we think our surroundings really are, while if I refer to a universe (or a model universe) I am talking about something that may be an interesting idea but is not necessarily believed to describe the way things really are.

the Big Bang—perhaps there was no "before"! But that possibility, as we shall see, doesn't stop cosmologists from speculating on what *might* have preceded the Big Bang.

MODERN COSMOLOGY AND OLBERS'S PARADOX

Theoretical cosmology—the study of the origin and evolution of the Universe—began in its modern form in the 1920s, following Einstein's publication in 1917 of a scientific paper which described how the equations of general relativity could be applied to describe the behavior of space-time on the grand scale. Cosmology really got off the ground then because at the same time large new telescopes were developed, which revealed from direct measurements the fact of the expansion of the Universe. But in a sense the theoretical developments, at least, were a century late—for simple evidence that was widely publicized in the 1820s could have been enough to show that the Universe is expanding, and if this interpretation had been made at the time there could have been some very interesting nineteenth-century cosmological theories, predating relativity. Perhaps, however, it was inevitable that true progress could only occur when theory and observation were working together and pointing in the same direction. Even so, it is surely worth taking a brief look at the evidence, requiring no more observational equipment than the human eye, which is alone enough to hint strongly at the expansion of the Universe, and therefore to imply that there was a Big Bang of creation in which the expansion began.

The German astronomer Heinrich Olbers published his paradox in 1826, and it is known by his name to this day, although in fact he was not the first person to speculate on the significance of the dark night sky. A couple of centuries ago, the few people who thought about the nature of the Universe for the most part accepted that it must be, on the grand scale, uniform, unchanging, and static. This was as much doctrine as anything else; the idea of change in the Universe simply was not fashionable. But as Olbers and others realized, a uniform, unchanging, and static universe does not fit in with the very simple observation that the sky gets dark at night.

A

B

C

The astronomers' "space probes." Observatories such as the Lick Observatory (A) are sited on mountaintops high above the clouds; inside the domes, (B) modern telescopes are festooned with electronic apparatuses to squeeze information out of the light arriving at the "business end." (Both pictures courtesy of the Lick Observatory.) Where actual photographs of the sky are needed, a special camera called a Schmidt Telescope is used (C, courtesy of UK Science Research Council), while the great dish at Jodrell Bank, England, has become symbolic of the modern science of radio astronomy (D).

D

This paradox can be presented both in simple, commonsense terms and in mathematical language. In everyday terms, if the Universe is infinite—if "up" extends forever—and it is full of stars (or galaxies), then in any direction you look your line of sight must meet up with the brightly shining surface of a star. So the whole night sky should be a blaze of light. The mathematics tells the same story, if anything, more strongly. To get some understanding of the Universe as a whole (or, in this case, the unchanging model universe of Olbers and his contemporaries) we can look at one typical part of it and extend the general features of that part to the whole. From our viewpoint here on Earth, a typical part of our surroundings in space, assuming that the surroundings are the same in all directions (isotropic), is a thin spherical "shell" of space with us at the center. The skin of a balloon makes a roughly spherical shell around its center; the atmosphere of the Earth is a spherical shell on a bigger scale, but our model shells are on a much grander scale still, encompassing whole clusters of stars and galaxies.

If we imagine one typical thin shell, at a distance R from us, then the number of bright stars or galaxies it contains will simply be the average density of bright stars or galaxies in the universe multiplied by the volume of the shell. And it happens that the volume of such a shell is simply its thickness multiplied by the square of its radius, R^2. With the basic assumption of uniformity in the Universe, although some of those stars or galaxies will be brighter than average and some dimmer than average, we can say that the bright and dim ones will roughly cancel out, and treat each glowing object in the shell as having average brightness. The brightness of all the light from the stars and galaxies in the shell is just the brightness of an average star or galaxy multiplied by the number of stars or galaxies, so that the brightness of the shell also depends on R^2.

But this is not the end of the story. For us, at the center of the shell, the light will appear much fainter, just as the lights of a city on the horizon can appear much fainter than the beam of a torch close by. And we know from studying the behavior of light that the apparent brightness of a source is just its actual, close-up brightness, *divided* by the square of its distance, R^2. This is no coincidence—the reason why the same conversion factor that applies to spherical shells comes in again is because the light spreads out from a central source evenly in all directions, "filling" an ever-growing sphere. So the same kinds of numbers come in once again.

The simplest example is with two shells of radius R and $2R$. The further shell contains four times as many galaxies (number proportional to R^2), but each one is one-fourth as bright as its nearby counterparts (brightness proportional to $1/R^2$). So each shell contributes exactly as much perceived light.

The importance of this for Olbers's paradox is crucial. For it turns out that the brightness of *any* shell of space around us, as seen from Earth at the center of the shell, is independent of the radius—the R^2's cancel out of the equation. In a uniform, static, and unchanging universe every small volume around us contributes as much light to the glow of the sky as every other. In an infinitely large universe, all of the separate bright shells would add up to produce an infinitely bright sky, which is clearly nonsense—hence the paradox. And what this tells us is that one, at least, of Olbers's assumptions must be wrong, as a description of the real Universe.

Perhaps the Universe is really quite small, with a definite "edge" beyond which there are no more stars and galaxies to contribute to the brightness of the night sky. That would resolve the paradox—but with modern telescopes it is possible to see a very long way out into the Universe, and certainly further than where the edge would have to be if this is indeed the best resolution of the paradox. Alternatively, the Universe might not be the same everywhere—if the galaxies far away are dim, then they won't contribute much light after all. But this is really the same explanation as having an edge to the Universe, except that instead of a sharp edge there is a gradual fading away. And, again, we can see far enough to be sure that although there are differences between galaxies nearby and those far away (more of this later in this chapter), the differences are not of the right kind to remove the paradox.

THE EXPANDING UNIVERSE

That leaves one alternative. The Universe *as a whole* must be changing, and in particular it must be expanding. In this case, the light from distant sources is weakened not just in proportion to R^2, but by an extra factor to "fill up" the extra space being created in the expanding Universe. The light is literally stretched on its journey, to longer wavelengths corresponding to the red end of the optical spectrum—a red shift. No energy has been lost in the process, but the available energy has been spread thinner.

Now, this is a very neat explanation of the nature of the real Universe, which uses a few very simple observations and one simple, but powerful, piece of theoretical reasoning to determine probably the most basic fact of life—that the Universe expands.* Curiously, though, this is not the way that cosmologists arrived at the discovery historically. The red shift in the light from distant galaxies was discovered, much to the embarrassment of the theorists, when they still believed (in spite of Olbers's Paradox) that the Universe was unchanging. Only after that discovery brought down the card house of their then current theories did they begin to take the expanding Universe model seriously.

So theoretical cosmology really began, as a scientific discipline, in 1917, when Einstein published a paper showing how the equations of General Relativity can be applied to describe the behavior of the whole Universe. What the equations actually describe is the geometry of space-time, and although they had been used with great success in tackling relatively "local" problems (the most famous, perhaps, being the prediction that light from a star would be bent as it passed near the Sun, and that this bending effect could be observed during an eclipse, which it duly was), Einstein naturally wanted to apply them to the description of *all* of space-time, the whole Universe. He tried to do this in line with the conventional wisdom of the time, which still (only just over half a

---

* Some people worry about the idea of an infinitely large Universe expanding, asking "what is bigger than infinity?" The answer is "more infinity," and the expansion is not just a mathematical trick since the expansion has real, physical effects. What happens is that the density of the Universe decreases—galaxies spread further apart—as well, of course, as the stretching of light and other electromagnetic radiation.

century ago!) held to the model of an isotropic, homogeneous, and static universe. But he failed. He found he could keep isotropy and homogeneity without difficulty, but the simplest models built using the equations of relativity were nonstatic. He always came up with either expanding universes or contracting ones, but never ones which stood still. In order to make his equations fit in with the way astronomers believed the Universe to be, Einstein had to introduce an extra factor, called the "cosmological constant." Quite frankly, he fiddled the equations. And in later life Einstein was to comment that this fiddle was the biggest scientific mistake he ever made.

But the basis for the fiddle soon disappeared, even though the cosmological constant was to have lingering effects on the theory of cosmology until the 1960s. During the 1920s, both theoretical cosmology and observational cosmology really began to get off the ground. The theory was improved by the Russian mathematician Alexander Friedmann, who produced what became the standard solutions of Einstein's equations in 1922. These Friedmann models—which remain the basis of mathematical cosmology to this day—describe the behavior of isotropic and homogeneous model universes, with or without cosmological constants, and they come in two basic types. The distinction between the two families of model universes depends on only one physical property—one parameter—the density of matter in the universe.

All of these models (leaving aside the mistake of the cosmological constant) involve expansion or contraction, or both. One very bizarre version starts out infinitely huge, with matter spread very thin, and squashes down into an infinitely dense mathematical point, a singularity. That clearly does not represent the real Universe, which we know is expanding, not contracting. But the other two basic families can each be fitted to the observations we make of the Universe about us.

Both *start* from a singularity and expand away into states of less density, which means today galaxies (more precisely, *clusters* of galaxies) moving away from one another, carried by the stretching fabric of space-time. But depending on just how much matter there is in the Universe, the equations tell us that either this expansion will continue literally forever, or the gravitational attraction of all the matter will one day first halt the expansion, and then turn it around into a collapsing universe. A universe that expands forever is dubbed, for obvious reasons, "open"; equally obviously, the generic name for the family of universes which recollapse is "closed." And you can draw an analogy with the open or closed situation for a universe and the critical escape velocity of a rocket or other projectile departing from the Earth.

As a projectile struggles up out of the Earth's gravitational field, it loses energy, which means that it slows down. Provided it is going fast enough to start with, it can escape from the Earth's gravitational influence altogether; but if it goes slower than this escape velocity it must lose all its energy and fall back to the ground. For a planet with more mass, but the same size as the Earth, a rocket starting off from the surface would need more energy—a bigger initial velocity—to escape. So the question is, Did the Big Bang set the expanding Universe off with a big enough initial impetus to overcome the "escape velocity" corresponding to all the mass of the Universe put together?*

As the Friedmann models make clear, if the density of matter in the Universe is less than some critical amount, then the Universe is open and must expand forever—it is infinitely big, and always was, even just after the Big Bang when its density was also very big. But if the density of matter today is greater than the critical amount, then the Universe is bent around on itself and closed—it is finite but unbounded—rather like the surface of a sphere. A spherical surface—the surface of the Earth, say—is finite in extent because it has a certain finite area. But it is unbounded because there are no edges; if you keep traveling across the surface in a straight line, you don't come to the edge of the world, but end up, eventually, back where you started. The difference is that the surface of the Earth is two-dimensional, wrapped around the third

---

* The real situation is a little more subtle than this. Einstein's equations describe how a gravitational field "bends" space-time, and the question becomes one of whether or not the self-gravity of the whole Universe is enough to bend space-time right around on itself, making the Universe closed, or whether it is left curved but still open. The escape velocity analogy falls down if you try to apply it to an infinite universe, because the implication is that there might be infinite mass and therefore an infinite escape velocity; but the analogy still helps, I find, in getting an intuitive feel for what is going on—very few people have any intuitive feel for the concept of bent space-time!

dimension of space. The space of the Universe is three-dimensional (and space-time is four-dimensional!) so that to make a closed universe there must, in some sense, be at least one "extra" dimension through which the wrapping around takes place. This, again, is an awkward concept unless you happen to be a mathematician, but there it is. Any awkwardness, however, is certainly compensated for by the fact that a finite but unbounded universe has a definite size, and we don't have to worry about the meaning of infinity. (But remember the size of the real Universe is very big, and certainly so big that, without the expansion, Olbers's paradox would still apply. The difference between "very big" and "infinite" may not be that much, in practical terms for you and me, even though it makes the mathematics more accessible.)

But a closed universe shares one important property with a planetary surface—if you set off in a straight line across space you will eventually get back to your starting point, from the other side, having circumnavigated the entire universe.

For our own Universe, the critical density of matter which would be enough to halt the collapse is now about 10^{-30} grams per cubic centimeter (a decimal point followed by 30 zeros before the 1). When the Universe was smaller, and denser, it was expanding faster (think of the rocket, shooting out into space and gradually slowing down), so this figure applies to the Universe as we see it now, with its observed rate of expansion. If we had been observing thousands of millions of years ago, we would have seen a more rapidly expanding and more dense universe, but because the rate at which the expansion slows down (the deceleration parameter) depends on the density at each cosmic instant (or epoch), if the amount of matter we see now is enough to close the Universe then we can be sure that the closure has been fixed permanently since the beginning, the Big Bang itself.

Once Friedmann's models had shown the critical importance of this relationship between the deceleration parameter and the density of the Universe, it was the turn of the observers to make the vital tests—first, to get some understanding of just how rapidly the Universe is expanding, and secondly to measure, somehow, the density of matter in the Universe, at least in our immediate cosmic neighborhood. Neither task was easy—and although 10^{-30} gm per

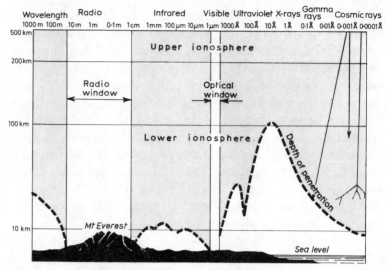

Observations made from the ground are limited by the extent to which radiation penetrates the Earth's atmosphere. Only the optical "window" was accessible to astronomers such as Hubble, who first worked out the red shift/distance relation. And only the radio "window" has provided a new view of the Universe since Hubble's time. But by raising instruments above the obscuring layers of the atmosphere in balloons, rockets, and satellites the astronomers of the 1980s are able to make observations of the Universe across the entire electromagnetic spectrum.

cu cm might seem a rather small density, the simplest calculations (guessing the mass of a galaxy, counting the number of galaxies in a large volume of space near us, and working out the mass density) soon hinted that the real Universe sits rather close to the dividing line between open and closed, making accurate measurements all the more important in order to find out on just which side of the line it does sit.

THE NATURE OF GALAXIES

The crucial observations which began to show that the Universe is expanding were being made in the same decade, 1910–1920, that Einstein was beginning to apply his relativistic equations to a description of the whole Universe. But it was not until well into the 1920s that any real understanding of the nature of the expansion began to emerge from further observations. Even at the beginning of the twentieth century, our own Milky Way Galaxy—the mass of stars we see in the night sky—was thought to represent all of the Universe. Various nonstellar objects had been discovered, and dubbed nebulae, but at first these were thought to be glowing clouds of gas between the stars. Some, indeed, turned out to be just that; but many others turned out to be whole galaxies in their own right, collections of stars like our Milky Way, but so far away that they appeared just as small patches of light even on photographic plates taken with modern telescopes.

One galaxy, like our own, may contain thousands of millions of stars. But in a dramatic new vision, the discovery of many, many other galaxies scattered across space beyond our own turned even this impressive collection of matter into a cosmic backwater. This was probably the greatest step in the shift away from the old view of the Earth as the center of the Universe. Over the space of only three or four centuries, this anthropocentric view had been replaced by the idea that the Earth is just one insignificant planet circling the Sun, then by the realization that the Sun is just one insignificant star among the myriads that make up the Galaxy, and now by the evidence that our Galaxy is but one insignificant stellar island among thousands of millions scattered across the sea of space.

Even before the extragalactic nature of many nebulae had been established, astronomers studying their spectra had found clear evidence of their motion relative to us. Starlight carries with it a "signature" characteristic of the chemical elements present in the atmosphere of the star, notably hydrogen and helium. This signature is in the form of patterns of lines in the spectrum of light, patterns which are easily seen with the aid of a spectroscope, which splits light up into its component colors, and which occur at precisely defined wavelengths and with the uniqueness of fingerprints. If an object moves toward us, however, any light we receive from it is compressed in wavelength, and shifted toward the blue end of the spectrum; for objects moving away from us the opposite effect, a red shift, occurs. So when astronomers found that the light from nebulae contained lines which were shifted compared with the wavelengths at which they are produced in the laboratory, they deduced that the nebulae were moving relative to the Earth.

At first, there seemed no clear pattern to these motions. Some nebulae, such as the great spiral nebula in Andromeda, showed blue shifts in their spectra, indicating motion toward us. Others showed red shifts, indicating recession. But as more and more observations came in, a pattern emerged. First, the pioneering modern observers, led by American Edwin Hubble, showed that many nebulae are indeed galaxies in their own right, beyond the Milky Way. Then Hubble, almost single-handed, found that only a few (nearby) galaxies show blue shifts, while most of them are moving away from us, with a velocity of recession that is greater the further away they are. Finally, Hubble formulated the rule now known as Hubble's law: the red shift in the light from a distant galaxy is directly proportional to its distance from us.

The crucial observations were made with a succession of giant telescopes built on mountains in California. The 60-inch reflector at Mount Wilson, completed in 1908, was a giant in its day, and was followed by the 100-inch at the same observatory in 1917 and the 200-inch at Mount Palomar in 1948. Hubble worked with all of these instruments, and from 1919 onward, at the Mount Wilson observatory, he showed first that the external nebulae could be resolved into individual stars (at least, those nearest to us could) and then estimated the distance to the Andromeda galaxy (our nearest large neighbor) and others.

This distance measurement is a saga in itself, being built up from a whole chain of evidence. Direct measurements of the distances to stars can only be made for our very nearest neighbors, using techniques which depend on the way they seem to move around the sky as the Earth moves around the Sun in its orbit— these are parallax effects, rather like the way nearby lampposts seem to move fast when viewed from the window of a moving train or car, while distant hills seem to move more sedately. Indeed, the parallax technique gives its name to the basic astronomical yardstick, the "parallax second of arc," or parsec. If a star was just far enough away from us that it seemed to move by one second of arc when the Earth moved across a distance equivalent to the distance from Earth to the Sun, it would be at a distance of one parsec, and from that star the radius of the Earth's orbit would cover exactly one second of arc. This is a rather small angle— there are 360 degrees in a circle, sixty minutes of arc in each

RELATION BETWEEN RED-SHIFT AND DISTANCE
FOR EXTRAGALACTIC NEBULAE

CLUSTER NEBULA IN	DISTANCE IN LIGHT-YEARS	RED-SHIFTS

H+K

VIRGO — 7,500,000 — 750 MILES PER SECOND

URSA MAJOR — 100,000,000 — 9,300 MILES PER SECOND

CORONA BOREALIS — 130,000,000 — 13,400 MILES PER SECOND

BOOTES — 230,000,000 — 24,400 MILES PER SECOND

HYDRA — 350,000,000 — 38,000 MILES PER SECOND

Red-shifts are expressed as velocities, $c\,d\lambda/\lambda$.
Arrows indicate shift for calcium lines H and K.
One light-year equals about 6 trillion miles,
or 6×10^{12} miles

The key to the scale of the Universe is provided by the red shift. Features in the spectrum of light from distant galaxies are shifted to the red end of the spectrum, indicating a velocity of recession. This velocity is proportional to the distance of the source from us—so the red shift provides a direct measure of distance across the Universe. In these pictures, from the Hale Observatories, the distinctive double dark line in the spectrum, marked by an arrow, is seen further to the right—further to the red—for more distant sources. (*Photograph from the Hale Observatories*) degree, and sixty seconds of arc in each minute. At a distance of one parsec, the whole of the Earth's orbit around the Sun would fill just two seconds of arc on the sky—yet such a distance is so local on the cosmic scale that there are no other stars within one parsec of the Sun!

To relate this cosmic distance scale to everyday yardsticks, it's convenient to use a unit beloved of science fiction writers but seldom used by astronomers, the light-year. This has nothing to do with time, but is the distance traveled by light in one year. At a speed of some 3×10^{10} (3 followed by 10 zeroes) centimeters every *second*, light covers 9.4607×10^{12} kilometers every year—63,240 times the average distance from the Earth to the Sun. A parsec is just over 3.25 light-years, and the nearest star to the Sun is 4.3 light-years away. Astronomical distances are most commonly given in thousands of parsecs (kiloparsecs, kpc) or millions of parsecs (megaparsecs, Mpc).

Our own Milky Way Galaxy is made up of some 100,000 million stars spread across a disk 30 kpc (roughly 100,000 light-years) in diameter and just over half a kiloparsec (some 2,000 light-years) thick. In all this immensity, literally only a handful of stars are close enough for distances to be measured by parallax, but a related technique, the moving cluster method, pushes the range of measurements out just far enough for the next step in the chain to take over. A nearby group of stars (or star cluster) moving through space together appears from Earth to be converging on one point in space, just as the parallel lines of a railroad track appear to converge at a point on the horizon. By measuring the velocity of stars in the cluster (which may need observations over years as the stars slowly move across the sky) and identifying the convergent point, the distance to the cluster and to each individual star can be determined. And, fortunately for astronomy and cosmology, there is one very large cluster of stars near enough to us for this technique to work, but containing a rich variety of stars all essentially at the same distance from our Solar System.

This is the Hyades cluster, rather less than fifty parsecs away, a group of stars moving through space together at a speed of 43 km per second. The cluster contains about 200 stars spread over a distance of three or four parsecs, enough to give an idea of how a star's apparent brightness in the sky is related to its actual distance and to other properties such as its color. Now distance measurements begin to be put on a statistical basis, relating color, distance, and apparent brightness for as many stars as possible. And one crucial discovery emerges from the statistics—one class of stars, the Cepheid variables, go through pulsation cycles which depend only on how bright (that is, how hot) the individual star really is.

An individual Cepheid will first brighten, then dim, then brighten again, with a period of between two and forty days. But its exact period—say, 10½ days—depends on exactly how bright it is (on average). The reasons for this are well understood, in terms of the balance between nuclear reactions which keep the star hot (more of this in Chapter 3) and the gravitational pressure forces squeezing it into a ball. But what matters for the distance scale is that if you know how bright a star really is, then its distance can be found quickly and accurately from its apparent brightness in the sky, using the familiar inverse square law (apparent brightness = actual brightness/distance^2).

This is the key link in the chain. Cepheids can be picked out from the mass of stars right across the Milky Way, helping to give us an idea of the size of our own Galaxy. And, crucially, Edwin Hubble found Cepheids among the stars of the Andromeda galaxy and other nearby galaxies.

Across such distances, even the size of a galaxy begins to be a small-scale "local" effect, and the first estimates of the distances to extragalactic nebulae such as the Andromeda galaxy came from averaging out the distances indicated by the pulsations of different Cepheids among their stars. We now know that our *near* neighbor in Andromeda is at a distance of almost 700 kpc, so remote that we see it by light which left more than 2 million years ago; and the Cepheid indicator can be used out to distances of about 3 Mpc (10 million light-years), roughly 100 times the diameter of our own Galaxy. This provides the foothold on the ladder of truly cosmological distances, but in cosmological terms even the nearest 3 Mpc is still just our own backyard.

RED SHIFTS AND THE SCALE OF THE UNIVERSE

Getting back to the pioneering work of Hubble, breaking new ground by measuring distances to external galaxies in the 1920s, we can pick up the story of universal expansion from the Big Bang. Hubble soon noticed that galaxies are distributed evenly across the sky (an isotropic Universe) and that all galaxies except our nearest neighbors show a red shift in their spectral features (an expanding Universe). In 1929, he announced that the velocity of recession indicated by the red shift is proportional to the distance of a galaxy from us, the famous red shift/distance relation now known as Hubble's law. This is exactly the behavior needed to fit the Friedmann models, based on Einstein's equations, and also the only kind of universal expansion which does not require the Earth and our Galaxy to be in a special place. If you imagine an expanding balloon, dotted with paint spots, then *every* spot will "see" every other receding with velocity proportional to distance, and no spot could be identified as being at the center of the expansion. Hubble's law shows that the Universe is made the same way—there is no "center of expansion," but everything recedes from everything else. And this fits in with the other great overall property of the Universe, that it is not just the same in all directions (isotropic) but would look the same from anywhere within it (homogeneous). So there is indeed just one difference from the older cosmological view that had caused Olbers so much head scratching—not an isotropic, homogeneous, static universe, but a homogeneous, isotropic, *expanding* Universe.

By 1931, the observational evidence showed the validity of Hubble's law out to 30 Mpc and recession velocities in excess of 20,000 km per second, and since then all new observational improvements have confirmed the validity of this Einstein/Friedmann view of the Universe.*

* You may wonder how distances were measured out beyond the Cepheid limit in order to establish Hubble's law. The first guess assumed that all galaxies were equally bright, and estimated distance solely from apparent brightness. Then Hubble tried guessing that the brightest galaxy in each cluster of galaxies must be at the limit of possible brightness, and therefore all the brightest cluster members should have the same intrinsic brightness. This crude approach gave just enough evidence to confirm Hubble's law, so that now the red shift is accepted as the best measure of distance beyond our immediate cosmic neigh-

So at last, after following the historical story through two crucial decades, we can sum up the modern cosmological view of the Universe. Our whole Galaxy, with its 100,000 million stars, is seen as just one ordinary island of matter in the immensity of space-time, the true fabric of the Universe. There are many, many other galaxies, some bigger than ours and some smaller, stretching as far as we can see across the Universe, some at distances of millions of parsecs, judging from the red shift/distance relation. Galaxies tend to occur in clusters, anything from a few to hundreds of galaxies moving together, and clusters are moving apart from one another, carried on the stretching fabric of space-time, as the Universe expands.

This, in a nutshell, is the picture which tells us that there must have been a Big Bang. First, just by imagining the Universe "running backward" we can see that it has evolved from a state of greater density with clusters of galaxies closer together; go back long enough and everything must have been squeezed in one lump. And, secondly, the Einstein/Friedmann models which so successfully describe the state of the Universe today all begin from a Big Bang, an initial instant of creation when space-time/matter-energy burst out from a unique point, a mathematical singularity.

COSMIC MICROWAVES

Now, all of our understanding of nature breaks down at a singularity—that is really what the word means, "a place where the laws of physics as we know them no longer work." So strictly speaking we can't describe the evolution of our Universe, or of a model universe like the real Universe, from the instant of the Big Bang itself, time zero. But possibly the greatest triumph of scientific thought has been the achievement, over the 1960s and 1970s, of a self-consistent description of everything that happened after the first one hundred-thousandth of a second, right up to the present day, some 15,000 million years later! This remarkable achieve-

---

borhood. But remember that it all comes back in the end to those studies of Cepheid variables, and when astronomers refine the Cepheid distance scale, as they still do slightly from time to time as more and better observations are made, then our distance measures of all astronomical objects, right out to the most distant galaxies, have to be revised accordingly.

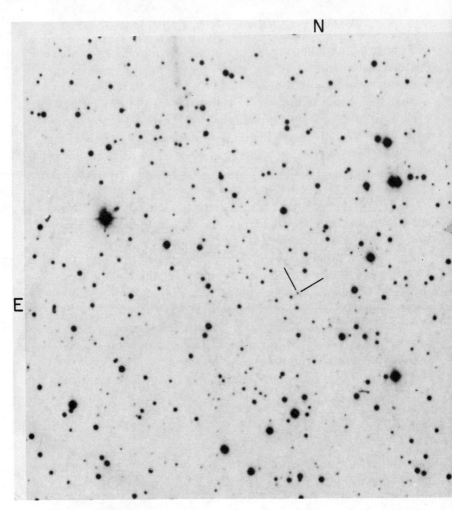

The insignificant fuzzy blob marked near the center of this picture, an enlargement of a print from the Palomar Sky Survey, is one of the most distant objects ever discovered in the Universe, the quasar OH 471. The measured red shift of 3.4 corresponds to a velocity of recession of more than 90 percent of the speed of light. (*Photograph supplied by R. F. Carswell*)

ment depends on two extra factors, which the present generation of scientists has contributed, to flesh out the basic Einstein/Friedmann models. First, we now know that the Big Bang was hot—the Friedmann models include universes which start out cold, as well as having other variables, and these models can now be eliminated. Secondly, high-energy physicists who study the nature and behavior of the so-called elementary particles, the building blocks from which more mundane particles like protons and neutrons are constructed, believe that they understand how such particles interact at the very high energies (high temperatures) and densities (pressures) that would have existed early in the history of a hot Big Bang universe.

The particle physics we'll have to take on faith; the studies themselves are rather esoteric, and their application to the Big Bang has been the subject of a whole book in its own right.* But a little digression may be in order to tell the tale of how we know the Universe was born in fire, as a *hot* Big Bang. For the story of the discovery and interpretation of the so-called cosmic microwave background radiation is a classic in its own right.

Arno Penzias and Robert Wilson would be the first to agree that they owe their discovery of the cosmic microwave background—the echo of the Big Bang itself—in large measure to a lucky accident. The extent of their luck—and the importance of the discovery—is reflected by the fact that they shared the Nobel Prize in physics for this work in 1978. The discovery had been made a decade and a half earlier, in 1964, when they were working at the Bell Laboratories with a sensitive radio antenna and receiver system designed for communications using the faint signals bounced off the *Echo* satellites.

The system had to be sensitive, since the *Echo* satellites were nothing more than huge balloons, inflated automatically in Earth orbit and covered with a film of metallic material to reflect radio waves. Signals beamed up from Earth stations would bounce off the satellites, with a very weak signal coming back down to Earth, where it could be detected thousands of miles away from the transmitter. Following the Bell company's policy of encouraging origi-

---

* Steven Weinberg, *The First Three Minutes* (see Bibliography).

nal scientific research, as well as research with clear commercial objectives,* Penzias and Wilson were using the antenna at Holmdel, New Jersey, to measure the background of radio "noise" coming from our Milky Way Galaxy in directions well away from the heart of the Milky Way itself.

In everyday terms, they were trying to measure the radio noise from the sky itself, since in these regions far from the center of our Galaxy any radio waves would be coming not from active individual stars but from the spread out contribution of the interaction of individual hydrogen atoms in space with the magnetic fields of the Galaxy. The expected noise would be so faint that a very sensitive antenna/receiver system would be needed to detect it. As well as the special construction of the twenty-foot horn antenna—looking like a giant ear trumpet—they used surpercooling techniques in which the power coming from the antenna was compared with the signal produced by a source kept cold with liquid helium. And after all this they found, in the spring of 1964, a much stronger signal† than anticipated, apparently coming evenly from all directions on the sky.

This posed a real puzzle. As the weeks and months wore on, the Bell team found no change in the signal with time of day or with the seasons, showing that, since the Earth rotates and moves around the Sun, constantly pointing the antenna to different parts of the sky, the signal must be genuinely isotropic. It was so strong, however, that if it was coming from the interstellar gas in our own Galaxy, then the Galaxy ought to be a radio beacon in the Universe; yet other galaxies like our own (especially our near neighbor in Andromeda) showed no sign of such strong radio emission at the wavelength of just over seven centimeters where the noise was detected.

It looked, in fact, as if this signal—strong compared with what Penzias and Wilson had expected, but still weak enough by any conventional standards—might be an unexpected side effect of the antenna design. Maybe the noise came from no further away in space than the end of the antenna itself, an idea which looked at

---

* This isn't just altruism, but ensures lots of publicity for Bell, as their mention here shows!
† "Signal" here does not mean a message from Little Green Men in space. Any radio noise from natural or unnatural sources is dubbed a signal.

least plausible since they knew that pigeons had been roosting in its throat, liberally covering the inside with pigeon droppings—a substance whose electrical properties were certainly not allowed for in the design! The pigeons were removed, and early in 1965 the antenna was dismantled, painstakingly cleaned, and rebuilt. It made scarcely any difference; the cosmic signal was still there, still coming isotropically from all directions in space. It had to be a genuine cosmic microwave background signal.

Such a signal has one very important property. Anything at all—including pigeon droppings—radiates a radio noise due to the movement of electrons in the material. The higher the temperature (above absolute zero, −273°C, the temperature at which all thermal motion would stop), the more the electrons move about, and the stronger the radio noise produced. Physicists and radio engineers use the concept of an ideal source of thermal radio noise, a so-called black body which radiates perfectly at all wavelengths (it follows that it must also *absorb* electromagnetic radiation perfectly at all wavelengths, hence the name). The strength of any radio noise can be compared with the theoretical intensity of radiation from a black body to give a measure of the temperature equivalent of the source producing the radiation. And the radio noise found by Penzias and Wilson—the cosmic microwave background—was equivalent to the radiation from a black body with a temperature between 2.5 and 4.5 degrees above absolute zero (2.5 K to 4.5 K). This is certainly a weak signal. But if it really fills *all* of space (more accurately, all of space-time) then we are dealing with a very great deal of energy altogether. How was it produced? What did the signal mean? The answers to these questions came very quickly once news of the discovery leaked out on the astrophysicists' grapevine, for in the same state of New Jersey, at Princeton University, a team was actually building an antenna specifically to test for the existence of a cosmic microwave background, whose existence had been predicted—unknown to Penzias and Wilson—by Princeton theorist P. J. E. Peebles!

Peebles had been working with a senior physicist at Princeton, Robert Dicke, who had pointed out to him that if the Universe has indeed evolved from a more compact and hot dense state, then it might be possible to detect leftover radiation from the hot phase

A

Arno Penzias and Robert Wilson (A), photographed with the horn antenna (B) with which they discovered the cosmic microwave background radiation, the echo of the Big Bang itself. *Credit: Bell Labs*

B

still permeating space today. When Peebles put the numbers into the appropriate calculations, he found that, indeed, a very strong flood of radiation seemed to be necessary if the Big Bang model really applied to the Universe we live in. The crucial point is that most of the matter in the Universe is hydrogen, the simplest element of all; yet at very high densities, equivalent to the explosion of the Big Bang away from a singularity, the basic particles (protons, neutrons, and electrons) would be so closely packed that they would tend to fuse rapidly into much more heavy elements. The only simple way to avoid this is if the very dense young Universe was filled not just with matter but with intense (that is, hot) radiation, not just radio waves or even light but radiation more powerful than X and gamma rays and the like, radiation with a huge black body temperature which blasted the nuclei of heavy elements apart as quickly as they were formed, and then cooled as the Universe expanded, leaving the hydrogen and a little helium to make up the stars and galaxies we see today.

The cooling of the original intense radiation is crucial. From one point of view, this is just another manifestation of the red shift, the very short wavelength radiation of the early Big Bang being stretched to fill the whole expanding Universe, and stretched ever longer (that is, weaker) as time goes by. From another point of view, the change is one of energy *density*, not total energy. Imagine a box divided exactly in half by a partition, on one side of which there is some gas and on the other a vacuum. If the partition is removed, gas will fill the whole box, but the amount of gas present in each cubic centimeter will be just half the amount that was present in each cubic centimeter of the gas-filled half before the partition was removed. Just the same effect would happen if one-half of the box had been filled with radiation to start with, although the "experiment" cannot now really be carried out and is entirely in the mind. Imagine our box—a black body—half full of radiation, and then with the barrier removed. The radiation spreads to fill the entire box, and the energy density falls. This weakening of the energy density is exactly equivalent to a fall in the black body temperature of the radiation. So, by analogy, the expansion of the Universe after the Big Bang should "cool off" the original radiation. And when Peebles put the first approximate numbers into the calculation he found that, in order to explain why only hydrogen and helium were left after the Big Bang, there should still be a leftover background radiation in the Universe today, with a temperature of something under 10 K. P. G. Roll and D. T. Wilkinson, working at Princeton under the guidance of Dicke, were building a system to look for this background predicted by Peebles when they heard of Penzias and Wilson's discovery! *

In the end, both teams published their initial findings on the cosmic background in the same issue of the *Astrophysical Journal*, one paper authored by Penzias and Wilson and the other by Dicke, Peebles, Roll, and Wilkinson. It does seem a little harsh that the team of experimenters and theorists who together predicted the presence of the radiation, found it, and explained its origin should end up with no more than a footnote in the history books, while the team that found the radiation by accident and couldn't explain what it was got half a Nobel Prize. But there is no denying that the discovery itself deserved such recognition.

Over the years since 1965, many observations across a wide range of wavelengths have confirmed the nature of the cosmic background, and set its temperature equivalent close to 2.7 K, at the bottom end of the range originally estimated by Penzias and Wilson. Now, in the 1980s, astrophysicists attempt to discern evidence of very subtle properties of the Universe from the tiny variations of the radiation away from a "perfect" black body spectrum—important work, but very much fine tuning compared with the literally cosmic significance of the background radiation's mere existence. For, having taken the exact temperature of the Universe today, we can wind the clock back to the beginning, in-

---

* The subtleties and ironies of the story actually go back even further. In the mid-1940s several people had pondered on the implications of a "hot Big Bang," and Dicke himself had been a member of a group at MIT which even then estimated the temperature of any residual radiation as below about 20 K. It seems he had forgotten this himself (except, perhaps, subconsciously?) when he set Peebles on the road to his own prediction two decades later! In the interim, other people had made similar calculations, but no one had made what seems with hindsight the obvious step of actually looking for the radiation, which could certainly have been detected with 1950's radio telescopes. It seems that the theorists themselves did not really believe, deep down inside, that their simple calculations actually applied to the real Universe out there. The discovery made by Penzias and Wilson was as important in this regard—showing theorists that they were indeed describing the *real* Universe—as it was in simple scientific terms.

creasing the temperature appropriately as the "box" of the Universe is compressed and energy density increases, and then start out from just after the Big Bang to the present day. In a sense, all of what has gone before in this book was preamble—necessary scene setting to explain how we can so confidently describe events close to the very beginning of space-time as we know it. But now, I hope after convincing you that cosmologists really do know what they are talking about, this is where the story *really* begins.

THE BIG BANG

The very beginning would be at time zero, when the density of the Universe was infinite. But we can't cope with singularities and infinities, so even though we may suspect that there really was a singularity at the beginning of space-time, we begin the mathematical description of the origin of our Universe from a time when the density of the Universe was huge but finite, and its temperature was correspondingly huge but also finite. How the very dense, very hot Universe came into being we cannot, in all honesty, say. We can, however, say how it got from being very hot and very dense to being almost empty and very cold, the state it is in today. The modern cosmological world view begins at a time when the Universe has cooled to "only" a million million degrees (10^{12} K), at a time one hundred-thousandth of a second (10^{-5} s) after the instant of creation. Even under these extreme conditions, the laws of physics deduced here on Earth can be applied to produce the self-consistent story of everything that has happened since.

At a temperature of 10^{12} K, particles and radiation are interchangeable; the energy density of black body radiation at such temperatures is so great that it can literally produce pairs of particles like protons and electrons, not out of thin air but out of thick radiation. The process follows two key rules, now very familiar to physicists not just from theory but from the experiments they perform using particle accelerators—cyclotrons, synchrotrons, and the like. First, the equation which so many people have seen that it has become a cliché (even if not everyone knows what it means), $E = mc^2$. This sets down Einstein's discovery that mass (m) and energy (E) are interchangeable, linked by a conversion factor of the square of the speed of light (c^2). With the speed of light so huge, 3×10^{10} cm per second, that means that the energy equivalent of a gram of matter is absolutely enormous; the energy equivalent of a tiny proton, weighing in at just 1.6×10^{-24} gm, is correspondingly tiny, and this is why individual protons can be created from the energy of very hot radiation, but whole grams of matter cannot be made in job lots in this way. The energy from the radiation is described by the other equation, $E = h\nu$, where ν is the frequency of the radiation (the inverse of its wavelength, so that the shorter the wavelength the bigger the frequency and the more energetic the radiation) and h is another constant, Planck's constant, some 6.6×10^{-27} erg seconds. Here, higher black body temperature corresponds to bigger ν, which means greater available energy E to make particles of mass m. In other words, the higher the temperature the bigger (i.e., more massive) the particles you can make.

So one hundred-thousandth of a second after the beginning, the Universe was a seething mass of particles and radiation, a swirling soup in which pairs of particles were constantly being created from radiation, and constantly being destroyed and turned back into radiation. For it happens that the way the Universe is constructed—obeying the laws of physics we have deduced from experiment—an energetic packet of radiation (a photon) cannot make just one electron, or one proton, or whatever. The creation always produces a pair, one particle and its "antiparticle" counterpart.

Antimatter is a form of matter which has almost mirror image properties compared with the matter which makes up you, me, and almost all of the Universe. Taking just one obvious property, electric charge: the charge on an electron is negative, while its counterpart, the positron, has exactly the same amount of charge but in the positive form. When a particle meets its antiparticle, the result is annihilation in a burst of radiation, mass being converted totally into energy. So while radiation energy was being turned into particle pairs, particle pairs were being turned back into radiation in the early Universe. Overall, though, the total mass/energy of the whole system—the whole Universe—was constant. For every E/c^2 of mass created or destroyed, an exactly equivalent E/h of radiation is always destroyed or created.

Things began to get more orderly as the Universe expanded and the temperature dropped to 10^{11} K, still within the first tenth of a second (0.1 s) of the Universe's life. At this temperature the exotic particles with high masses could no longer be created, and any weirdos that had been produced soon met their antiparticle counterparts and annihilated. Only electron/positron pairs and the massless neutrino/antineutrino pairs were light enough to have a continuing involvement in the matter/radiation balance. How big was the Universe then? The question is a natural one but may be meaningless. If the Universe is open and able to expand forever, then it is infinite in size *and always has been.* In such a universe, infinite in extent even just after the Big Bang when it was very hot and very dense, temperature and density provide the only sensible indicators of its evolution, with expansion producing a cooler, thinner but still infinite universe. On the other hand, if the Universe is finite and closed, the question can be answered—but the answer is a puzzling one! In this picture, the circumference of the Universe has grown from zero, and is inversely proportional to the black body temperature—the temperature of the background radiation—at any time. After 0.1 s, at 10^{11} K, the circumference would have been about four light-years, just over a parsec, on this model.

Here is a curiosity indeed. It is deeply ingrained in the heart of relativity theory, the theory on which these universe models are built, that the speed of light is both an absolute constant (in vacuum) and the ultimate "speed limit." To start from zero volume and produce something four light-years around in only 0.1 s seems clearly to violate that speed limit. But the point is that the limit applies to matter moving through space-time, while the expansion of the Universe involves the evolutionary change of space-time itself. Remember, matter doesn't flow through space-time in the expansion, it is just carried along for the ride; the red shift we see in the light from distant galaxies doesn't indicate that those galaxies are moving through space away from us, but that the space-time between us and them is stretching. In a very real sense, those galaxies are stationary relative to space-time in their locality; and it is indeed possible that there may be galaxies so far away from us across the fabric of space-time that light from them can never reach us, since the space-time in between is stretching faster than the light can cross it, giving an effect as if, from our point of view, they are receding faster than the speed of light.

Either way, open or closed, the Universe is clearly a complicated place where "common sense" does not always apply! But in either picture the continuing evolutionary story can be told in terms of decreasing density and temperature, not of "size" (whatever that means). The open or closed question, however, may have important repercussions in the distant future, as we shall see.

LEFTOVER MATTER

About 14 s after the Big Bang, the temperature of the Universe had dropped to around 3×10^9 K, and the weakening radiation no longer had the strength to create pairs of electrons and positrons. Most of the electrons met up with their opposite numbers and annihilated; the great days of mass/energy interchange were over, and the Universe became a lot quieter and a lot emptier. But not quite empty. For, by some quirk of nature we do not yet understand—and may never—a few "extra" electrons were left over, along with protons, at least in our corner of the Universe. Enough matter was left to form the stars, galaxies, and planets around us, virtually everything that is important in the Universe from the viewpoint of life as we know it. If the matter/energy balance had acted entirely symmetrically, producing no excess of either matter or antimatter, everything might have annihilated, leaving a universe expanding like ours, filled with background radiation like ours, but containing no matter at all. Just possibly the amount of matter in the Universe is balanced by an equal amount of antimatter, but the two have become separated in some mysterious way. Maybe some of the galaxies astronomers study through their telescopes are made of antimatter, with antimatter suns, antimatter planets, and even antimatter people. Or maybe some as yet unfathomed asymmetry just after the Big Bang produced a tiny surplus of matter to become the visible Universe we know. Either way, there are hints here of a very profound relationship between the existence of life and the precise details of the way in which our Universe is constructed—a theme to which I shall return.

But matter *was* leftover as the final decoupling from the radiation background took place, and from then on the two followed separate paths. The radiation had nothing left to play with, and simply cooled as the Universe expanded, all the way down to the 2.7 K echo of its former glory that we detect today. For matter, the adventure was just beginning. At around 10^9 K, some seventy times the temperature in the heart of the Sun today, many protons and neutrons fused into helium nuclei, and as the cooling continued these later picked up electrons to become stable helium atoms. At the same time, the cooling allowed the remaining protons to pick up electron partners and form hydrogen atoms. By the end of the first four minutes after the Big Bang, 75 percent of the remaining mass of the Universe was in the form of hydrogen nuclei, and the rest had been processed into helium nuclei; it took a further 700,000 years for cooling to proceed to the point where electrons became bound to the nuclei to make atoms, at a temperature around 5,000 K. Strictly speaking, up to this time the radiation had still been able to interact with matter, since free electrons and protons, being electrically charged, do interact with radiation, even though no more annihilation/creation processes occurred after the first four minutes. So when we measure the background radiation today and find that, to a remarkable degree of precision, it is both isotropic and homogeneous, it is telling us something about the Universe as it was 700,000 years after the Big Bang, when radiation and matter finally ceased to interact. We can confidently say that since that epoch the Universe was homogeneous and isotropic; we cannot say for certain that it didn't reach this neat state from a more complicated earlier phase, but the evidence is entirely consistent with the view that the Universe has been isotropic and homogeneous—except for the curiosity that a little matter was left over when the mass/energy interchanges finished—ever since the Big Bang.

After the first 700,000 years, then, the story of the Universe is the story of matter—galaxies, stars, planets, and life. The age of the Universe then was rather less than one-twentieth of its age today; the black body temperature was equivalent to the temperature at the surface of the Sun today. Compared with the first four minutes, this was a feeble flicker of energy, with the story of creation virtually over. Looked at from here and now, that was the fiery furnace out of which we were born.

2 THE ORIGIN OF OUR GALAXY

After the fireball stage, the Universe began to look the way it is today, with matter concentrated into glowing lumps (stars) clustered together in material islands (galaxies) scattered across empty space. But the way in which galaxies formed, and the way they have subsequently evolved, depended on their inheritance from the fireball.

When we look at the night sky on a cloudless night (and away from the bright city lights!) we see an impressive display of bright stars. The display is so impressive, indeed, that it comes as something of a surprise to learn that only about 3,000 stars can be distinguished by the naked eye even on the darkest, moonless night. No matter how long you stand looking at the sky, you will never see more stars than this, since the human eye adapts to the conditions as far as it is able very quickly. But a photographic camera works quite differently from the human eye, and the longer its shutter is kept open the more light it gathers. So a long exposure photographic plate, taken with the aid of an astronomical camera, shows many more images than a short exposure plate, and the longer the exposure the fainter the objects that are recorded on the plate.

So it was from astrophotography, not the unaided observations of the human eye, that astronomers learned of the vast number of faint stars, and the very large number of faint, fuzzy "nebulae," quite different from stars, also revealed in astronomical photographs. This evidence suggests that there are thousands of millions of stars making up an island in space some thirty kiloparsecs across—our Galaxy. Roughly speaking, these are similar objects to our Sun (although there is a great range of stellar sizes and brightness), and they look so faint because they are, in round terms, tens of millions of times farther away from us than the Sun. Many of the fuzzy blobs, we now know, are whole galaxies of stars in their own right, thousands or millions of times farther away still, and we can only see them at all because they each contain hundreds or thousands of millions of stars. The distances to the galaxies—all but our very nearest neighbors—are so great that the light from a thousand million stars like the Sun appears as a faint, fuzzy patch on the sky, no brighter on a long exposure photographic plate than the dim image of a faint star.

As far as life on Earth is concerned, these fuzzy blobs on photographic plates might seem to be of very little significance. But it was from studies of these fuzzy blobs that Hubble and his contemporaries first produced convincing evidence that the Universe is expanding, with all that that implies for the Big Bang and the origin of everything. And if we want to understand our own Galaxy, and the place of the Sun and Solar System within our Galaxy, then it certainly helps to look at galaxies in general, just as it is easier to gain an understanding of the nature of trees in general by looking at a whole forest rather than just examining one specimen.

BRIGHT STARS IN A DARK UNIVERSE

The most striking feature of the Universe today is that it is a dark place which contains local lumps (galaxies) of brightly shining material. I have already discussed the implications of this in terms of Olbers's paradox, but it is also puzzling in another sense. One of the most fundamental patterns of behavior we find in the physical world is the tendency for things to run down and wear out. Things tend, to use the jargon term, toward a state of thermodynamic equilibrium. If we put a lump of ice in a cup of hot water, the ice melts while the water cools and we are left with a bland, in-between state. If we put bright, hot stars into a cold, dark universe, then the same pattern of behavior ought to result in the stars giving up their energy to (very slightly!) warm up the universe before they too become cold and dead and thermodynamic equilibrium is re-stored. In such a state, all the matter in the Universe would be at a temperature of just under 3 K, in equilibrium with the cosmic background radiation (this is turning Olbers's paradox on its head; you might argue, just as logically, that to produce thermodynamic equilibrium the background radiation ought to be as hot as a bright star, so that the sky is a blaze of light). The simplest expla-nation of the thermodynamic disequilibrium we see in the Uni-verse around us is that the Universe is, indeed, winding down. The Big Bang produced the disruption of equilibrium equivalent to dropping the ice cube in the hot water, and we live in a time when things are moving back toward equilibrium, but are still quite a way from it. In this picture, the very recent realization that at least 90 percent of the matter in the Universe is in the form of cold, dead stars is a very happy one, because it shows that by far the majority of the matter in the Universe *is* in equilibrium with the cold background radiation!

The bright galaxies—those fuzzy blobs that stop photographs of the depths of space from being completely black—are just the last flicker of disequilibrium as matter continues to adjust after the convulsion of the fireball. This important understanding gives us an appropriate new perspective, even when looking at the way galaxies were described in the days when astronomers only had op-tical observations to work with. Today, radio telescopes play as big a part in the study of galaxies as astrophotography does—and ob-servations across the electromagnetic spectrum at X-ray wave-lengths, in the ultraviolet and so on, are becoming increasingly important as more instruments are hoisted into space in satellites orbiting above the obscuring layers of the Earth's atmosphere.

Hubble originally classified the different kinds of galaxies visible on astronomical photographs according to their shapes—which was certainly a reasonable thing to do, since that was just about all the evidence he had to go on in the 1920s. His classification remains in use today, somewhat modified, and at the heart of it is the distinction between elliptical galaxies, which do indeed look like ellipses in astrophotographs, and spiral galaxies, which are made up of a central bulge of stars rather like a small elliptical galaxy and a surrounding disk of stars and other material through which spiral arms can be traced. Galaxies which fit neither category are called irregulars, and the spirals are divided into two subcategories, ordinary spirals, in which the spiral features can be traced out from the central bulge (the nucleus) and barred spirals, in which the spiral pattern seems to begin at the opposite ends of a bar of stars across the nucleus. Further subdivisions categorize the galaxies according to how tightly the spiral arms are wound, whether an elliptical is more nearly flat or more nearly spherical, and so on. But these differences are of no great significance. Nor is there any real significance in a classification made much later, dividing galaxies into "normal" galaxies (which are quiet and well-behaved) and "active" galaxies (which seem to be exploding or involved in other violent activity). It now seems that *all* galaxies pass through active phases during their lives and that therefore it is quite normal for a galaxy to be active!

Over the decades since Hubble's original classification, many astronomers have added subtleties to it, with subclassifications and sub-subclassifications. This is the kind of nit-picking attention to detail sometimes referred to as astronomical stamp collecting—adding new categories for the sake of it. And since the breakthrough in the late 1970s has made it possible to explain the variety of galaxies we observe in terms of the events of the post-fireball era and subsequent interactions, there is no need for this kind of repeated subdivision and categorizing. The important point which does have to be explained by the new model is that 75 percent of observed galaxies are spirals (including our own Galaxy), 20 percent are ellipticals, and 5 percent are irregulars. In addition, since there seem to be very few small spiral galaxies in our neighborhood, but plenty of very small ellipticals and irregulars, these observations probably do not give a fair picture of the balance of the Universe, and the best informed guesses ("guesstimates") place the balance of spirals and ellipticals quite close, with perhaps more ellipticals than spirals if all the tiny galaxies are included.

THE STRUCTURE OF GALAXIES

So much for the appearance of galaxies. What of their structure—what are they made of and how do their components interact? The most important property seems to be rotation. Just from their shape, spiral galaxies in particular seem to be fairly rapidly rotating objects, with the disk and spiral arms sweeping around the central nucleus. This is borne out by observations—even though it may take a couple of hundred million years for a galaxy like our own to turn once, so that there is no hope of watching it turn in a human lifetime. Once again, like the evidence of the recession of galaxies from one another, the observations depend on the Doppler shift in spectral light produced by motion. At its simplest, imagine a spiral galaxy seen edge on, rotating so that one side of the disk is moving toward us while the other side, across the nuclear bulge, is moving away. Clearly the rotation will show up as a Doppler blue shift on one side of the bulge and a red shift on the other.

For most galaxies, the situation is far from that simple. First, of course, the average red shift of the whole galaxy, caused by the expansion of the Universe, must be accounted for. And very few galaxies are conveniently aligned edge on to us! But the effects of the velocity of rotation in different parts of the galaxies can be unraveled even in nonideal cases, aided partly by one effect of the relativistic time dilation, which produces a red shift, depending only on velocity, in the light from a star moving directly *across* the field of view. All of this evidence shows that the disks of spiral galaxies do indeed rotate rapidly—and if "rapidly" seems a strong term for something that turns once in 200 million years or so, remember that the wheel that is turning is thirty kiloparsecs or so across. Spiral galaxies are, in fact, rotating just about as fast as they can without breaking up and flying apart—and this is exactly the sort of behavior that would be produced if the stars we can see were made from the material of a bigger cloud of gas which had collapsed under its own gravity into a more compact state. The collapse would produce a roughly balanced system, in which the

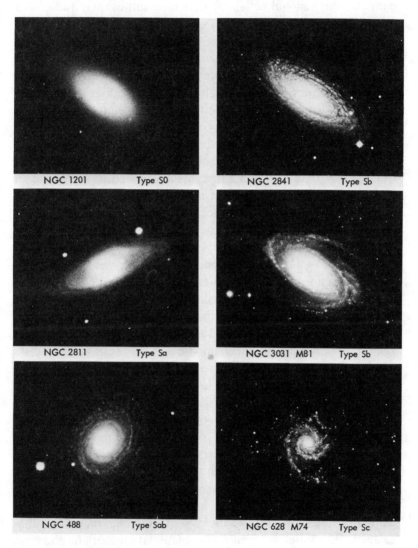

NGC 1201   Type S0

NGC 2841   Type Sb

NGC 2811   Type Sa

NGC 3031 M81   Type Sb

NGC 488   Type Sab

NGC 628 M74   Type Sc

The variety of spiral galaxies is shown in this montage of pictures from the Hale Observatories. (*Photograph from the Hale Observatories*) size of the galaxy produced depended on the rotation of the initial cloud; as the material collapses, the cloud spins faster, and gravity cannot pull the system into a more compact state than the rotation allows.

Velocity distributions can also be measured for elliptical galaxies, and they rotate much more slowly. This, again, is roughly what astronomers expect—if one effect of rotation is to flatten things into disks, then the guesstimate is that more rotund, elliptical galaxies are not flattened by rotation. But this isn't the end of the story, for measurements made during the 1970s, as techniques have improved, have shown that elliptical galaxies rotate *much* more slowly than the appropriate breakup speed for their size. It looks as if they have formed in a quite different way from spiral galaxies, since if they too had formed from collapsing gas clouds, why should the collapse have stopped while gravity still dominated over the effects of rotation?

This very new evidence encouraged astronomers a couple of years ago to think again about the shape of elliptical galaxies. They *look* rather like American footballs, or cigar shapes, and everyone from Hubble on had assumed that ellipticals really were like this (oblate ellipsoids). But it would be possible to get a pattern of galaxies that looked elliptical on the sky but were actually disks tilted away from us by different amounts (saucer-shaped, or prolate). Fortunately a very simple test distinguishes between the two. When we see a circular galaxy on the sky, we are either looking down the length of a fat cigar or right through a saucer-shaped galaxy. In the former case, we would be looking down a long column of stars, so circular galaxies should be bright; in the other case, we are looking through the thin disk, and circular galaxies should therefore appear relatively dim. And when astronomers checked their photographs they found that, indeed, elliptical galaxies that look circular on the sky are brighter than long thin galaxies. The ellipticals really are oblate (fat cigars), and there really does seem to be a fundamental difference between their rotation and the rotation of ellipticals, suggesting that there are two different ways to make galaxies.

There are other differences between spiral and elliptical galaxies. Ellipticals contain mainly old, red stars, while spirals seem to have two different "populations." The stars of the disk—like those near the Sun in our own Galaxy—are relatively young, and the spiral pattern is invariably marked by hot, blue stars; the stars of the central bulge, on the other hand, tend to be old stars, and red, rather like the stars of a typical elliptical. In a way, spirals look like ellipticals with spiral arms of young stars wrapped around them, or ellipticals look like spirals stripped of their disks. But the idea that a spiral might "lose" its disk and evolve into an elliptical doesn't fit the rotation evidence—nor does it account for the fact that although there are many small elliptical galaxies, the very biggest galaxies of all are invariably ellipticals. In among the young stars of the disk ("Population I"), spirals also contain a great deal of dark material, clouds of cool hydrogen gas plus dust (mainly carbon) and a scattering of molecules such as carbon monoxide, water, and formaldehyde. Ellipticals, with their "Population II" stars dominant, seem to contain very little interstellar material, except possibly for a trace of hydrogen.

Hydrogen and the other constituents of the interstellar medium can best be studied at radio wavelengths, and hydrogen in particular can be "mapped" at its characteristic wavelength of 21 cm. Deviations from the exact 21-cm wavelength are attributed to Doppler shifts, which affect all of the electromagnetic spectrum, not just visible light, so that radio astronomy provides a powerful additional tool for mapping the rotation curves of galaxies. It also provides scope to extend observations of galaxies out from the bright stars, since in most cases the hydrogen emission can be detected over a spread of space either side of the bright stars of the photographic galaxy. Although the hydrogen in spiral galaxies generally follows the spiral pattern, it may be detected twice as far out from the nucleus as any star of the disk. So the radio techniques give the best overall picture of the overall rotation curve of a galaxy—how the speed of rotation varies with distance from the center. And it is this evidence which, in the last half of the 1970s, clinched the argument that bright galaxies must be embedded in much bigger, more massive dark halos. In essence, the rotation curves show the dragging effect of material outside the bright disk of a spiral galaxy, material which doesn't shine and cannot be seen directly, but which interacts with the bright galaxy through gravity

NGC 4374

NGC 4564

Elliptical galaxies show less variety of shapes than spirals, simply ranging from spherical, like NGC 4374, to cigar or spindle shaped, like NGC 4564. Recent observations, depending on measuring the brightness seen "through" ellipticals at different points, have confirmed that this pattern really is due to differences in

NGC 4459

NGC 4473

shape and that a galaxy like NGC 4374 really is spherical, not a cigar-shaped galaxy viewed end on.

Montage prepared from UK Schmidt Telescope pictures supplied by the Royal Observatory, Edinburgh. Copyright © 1980

and makes its presence felt through the pattern of the rotation curve. This dark material is not cold gas, which would show up at 21 cm and would, in any case, soon get tugged into the central galaxy by gravity. It might be something exotic, like thousands of millions of tiny black holes scattered through space. But it can be explained, quite simply, as a mass of burned-out stars, left over from the first phase of star formation just after the fireball. This is one of the great triumphs of modern theoretical astrophysics—a model which explains how galaxies formed (in which the supermassive halos play a crucial part) and how elliptical galaxies have evolved from the spirals which came first. We can now take the story from the fireball up to the appearance of a galaxy like our own—and beyond—using the latest ideas which first appeared in this form in a scientific journal only in late 1978.*

The problem is to explain how hot, bright stars got scattered across the cold, dark Universe, when, according to the best modern understanding of the Big Bang origin of the Universe, the fireball itself was pretty uniform and very close to thermodynamic equilibrium. It turns out that everything depends on that one qualification "very close," for the miserable amount of bright matter that there is in the Universe today is just a tiny departure from equilibrium, built from tiny fluctuations in the balance between matter and radiation at the end of the fireball stage. Remember, too, that the presence of any matter at all is only the result of a tiny fluctuation away from equilibrium in the fireball; in a perfectly uniform universe all matter and antimatter would have annihilated. So we are now talking about the effects of a tiny fluctuation within what was already, on the cosmic scale, a tiny fluctuation. Galaxies like our own represent not just an afterthought of creation, but an afterthought of an afterthought! So it is all the more remarkable that creatures inhabiting one small corner of one single galaxy should be able to produce a coherent, self-consistent picture (model) of how it all happened.

---

* The journal was the *Monthly Notices of the Royal Astronomical Society*, vol. 183; the authors of the paper were S. D. M. White and M. J. Rees (see Bibliography).

Our own spiral Galaxy, the Milky Way system, would look something like this if we could view it from outside. The Sun is about two thirds of the way out from the center of our own Galaxy, near the edge of a spiral arm. (*Photograph of galaxy Messier 81 from the Hale Observatories*)

GALAXY FORMATION

The story of galaxy formation begins with fluctuations in the last period of the fireball era, when the Universe was filled with hot matter and hot radiation. The matter was what we would call an ionized gas, with negatively charged electrons and positively charged nuclei (chiefly single protons), still able to interact with radiation and not bound into electrically neutral atoms. The constant interactions between the charged particles and radiation kept the two forms of energy distributed smoothly, and the smoothness of the cosmic background radiation today, which has scarcely interacted with matter since that time, is a sign of the smoothness of the Universe then. But there must have been some fluctuations from perfect uniformity among the particles of the ionized gas. As they moved about at random, it must have happened that from time to time one spot would get more than its share, temporarily increasing in density, while somewhere else there would be a temporary deficit of particles, thinning out the cosmic soup. And radiation, too, under those conditions, is subject to the same kind of fluctuations in energy density. Uniform the Universe certainly was, over large spans of space and reasonable stretches of time. But it must have been bubbling with activity as, first in one place and then in another, denser or thinner pockets of mass or energy formed, only to be dissipated by matter-radiation interactions and re-formed into new patterns. The pattern of bright galaxies we see today is left over from the very last fluctuations of the fireball, the pattern that became frozen in as the Universe cooled to the point where nuclei claimed their quota of electrons to form electrically neutral atoms, and radiation and matter decoupled, leaving gravity as the dominant universal force.

In a mixture of ionized gas and radiation, there are two kinds of density fluctuation which can occur. If it is just a question of particles getting together briefly in a patch of increased density, then it is called an "isothermal" fluctuation (because there is no change in energy density, which corresponds to temperature); if, however, the fluctuation increases the local density of both matter and radiation, it is called "adiabatic." Both kinds of fluctuation must have occurred in the fireball. But the pattern of galaxies left over today very clearly shows that isothermal fluctuations dominated at the end.

Rather than trying to calculate the later evolution of a universe filled with both kinds of fluctuation, astronomers have looked at each kind and its consequences separately. Purely for historical reasons—because some very significant pioneering work on the subject was carried out by Academician Ya. B. Zel'dovich—Russian theorists have tended to concentrate on the growth of adiabatic fluctuations in an expanding universe, while Western theorists have looked in more detail at isothermal fluctuations.

The crucial difference between the two kinds of fluctuation is that whereas isothermal fluctuations (matter only) can be of any size, adiabatic fluctuations (matter plus radiative energy) can only survive and grow if they start bigger than a critical size. Small fluctuations are quickly smoothed away and damped out by the changes in energy density of the radiation. And the implication of this difference is that if isothermal fluctuations dominated in the era of recombination, then the Universe should be filled with many relatively small galaxies, clustered together into larger groups, which themselves cluster into superclusters, and so on in a continuing hierarchy. But if adiabatic fluctuations dominated in the real Universe, then the first aggregations of matter must have been many times more massive than even a cluster of galaxies as we see it today. In that picture, the original aggregations of matter would have been huge gas clouds, which then collapsed under gravity and broke up into smaller clouds, which in turn broke up into individual galaxies and stars. So the adiabatic fluctuations produce not a hierarchy of clusters (like a set of nested Russian dolls), but a uniform pattern of very similar superclusters side by side (like a box of toy soldiers).

To analyze the clustering patterns of galaxies in the real Universe requires a lot of patience and a good, fast computer to do the calculations. More than a million galaxies have now been located as members of one cluster or another, and the statistical calculations show very strong support for the isothermal (hierarchical) model. At the same time that these calculations were being made in the late 1970s, improved calculations of the physics of the recombination era were also being made, drawing on the improved cosmology of the 1970s which had been built on the discovery of the cosmic background radiation. These latest estimates suggest that, in fact, any adiabatic fluctuations forming at that time would be vastly bigger than even a supercluster of galaxies, and so stable that they would remain forever as huge gas clouds, never fragmenting into galaxies as we know them at all! So two lines of attack both point to the same conclusion: the distribution of matter in the Universe today results from isothermal fluctuations in the distribution of matter (local density fluctuations) in the last phase of the cosmic fireball, during recombination.

So far, so good. But how did galaxies like our own actually form from the primordial fluctuations? The complete story has still to be unraveled. But the outlines at least are now becoming clear. And the development of the new picture involves a major upheaval in astronomical circles, where a wealth of new evidence is coming in to show that many of our old ideas about galaxies—and by "old" I mean pre-1975—were quite simply wrong. Writing in 1980, conveniently at the start of a new decade, it is possible to say confidently that no book published in any previous decade, from academic text to popular account, gives a reliable picture of what galaxies are, let alone how they formed. For none of those accounts pays more than passing attention to the evidence, from rotation curve studies, that 90 percent of the mass of the Universe is in the form of dark stars, which by analogy with the two "populations" of bright stars in galaxies are now sometimes referred to as "Population III."

The White-Rees model, the first coherent attempt to produce an overall picture of the new understanding of galaxies, is based on the reasonable argument that just after recombination, when the overall density of the Universe was much higher than it is today, conditions were ideal for star formation and the first stars formed then in great numbers. These stars may have been very small, something intermediate between our Sun and the giant planet Jupiter, so that they never released much in the way of nuclear energy and never burned very brightly. Or they may have been very big—superstars, perhaps containing the matter of a million Suns—which burned their nuclear fuel quickly, scattered the "ashes" in the form of heavier elements built up from hydrogen

The way galaxies are clustered together in the Universe provides a clue to conditions in the hot fireball of matter and radiation out of which stars and galaxies formed. In the picture of the cluster in the direction of the constellation Hercules, every fuzzy blob represents a galaxy in its own right, comparable to our Milky Way system. Just below the center of the picture is an interesting pair of interacting spiral galaxies; tidal forces at work in such interactions, and head-on collisions between galaxies, turn some of the original spirals in clusters into ellipticals, stripped of their spiral arms. (*Photograph from the Hale Observatories*)

and helium by nuclear fusion, and left behind burned-out cinders, neutron stars or black holes. Either way, the dark stars left behind would have been distributed through the hierarchical clustering that was the heritage from the original isothermal fluctuations. And only 10 percent of the original hydrogen and helium gas was left to cool in the space between these early stars.

As the Universe continued to expand, moving the clusters of now dead stars apart from one another, the tiny remnant of old, cold gas must have sunk into the middle of each supergalaxy, sliding down the gravitational potential well and collapsing to form the bright galaxies that we see today, embedded deep within the real galaxies of Population III remnants. Any new theory must be tested by confrontation with observations, and the White-Rees theory stands up to such confrontation very well. First, and crucially important, this kind of model can be developed to explain why there are two kinds of bright galaxy, spiral and elliptical, visible in the Universe today. (Irregulars can be explained by almost any theory as the leftover bits and pieces of other processes; it is explaining *structured* galaxies that is difficult!)

When any cloud of material is compressed, there is a tendency for it to spin faster, to conserve angular momentum. The same thing happens to a spinning skater whose arms are tucked in; the more compact a rotating body becomes (however big it started) the faster it spins. So the cold gas falling into the center of a Population III galaxy must have ended up spinning rapidly, no matter how slow its rotation was originally. As such a collapsing cloud spins, it flattens; gravity can continue to pull material in down the "poles," but rotation holds matter out against the pull of gravity around the "equator." So something very like a spiral galaxy should emerge from such a collapsing gas cloud as it reaches the kind of density where star formation can begin.

Where do ellipticals come from? This is where the computers come in again, together with more observations of the velocity distribution of stars *within* typical elliptical galaxies (using, yet again, the Doppler analysis technique). The computer simulations show that when two spiral galaxies meet one another (as they will do from time to time as they move through a cluster; the expansion of the Universe carries clusters ever further from one another, but

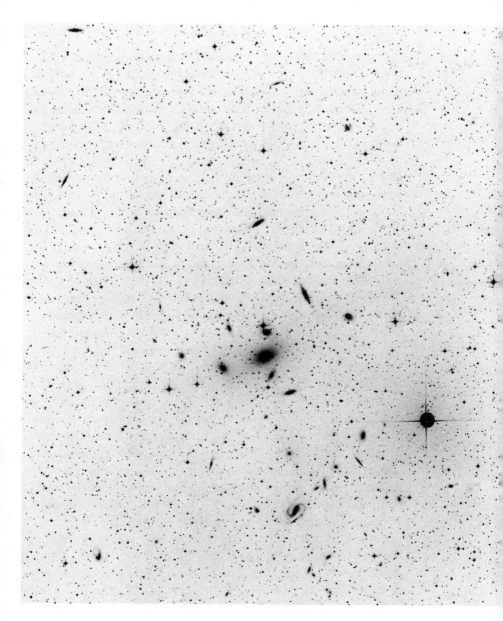

Another rich cluster of galaxies, printed as a negative (black on white) to show detail more clearly. Fuzzy blobs are galaxies in the cluster in the direction of Pavo; round dots are stars in the foreground in our Galaxy, and the "spikes" sticking out of them are produced by the camera and should be ignored. This photograph was taken with the UK Schmidt Telescope (Royal Observatory, Edinburgh), shown on page 9.

this does not affect the movement of galaxies within clusters, bound together by gravity, including the gravity of all those dark Population III stars), then gravitational interactions (tides) greatly change them. The outer disks are affected most, and even close encounters between galaxies can rip the disks apart, scattering trails of stars across a sweep of space many times greater than the diameter of a bright galaxy. Such interacting systems now seem to be able to explain the variety of very peculiar looking galaxies that have been discovered here and there on astronomical photographs; and this achievement alone represents a triumphant product of the combination of skillful human programming and high-speed computers—it is necessary to simulate *two* galaxies in the computer, then to simulate the tidal interactions as they pass one another, with the computer drawing sketches of the ever-changing patterns of stars in the interacting galaxies as the close passage proceeds.

But close passages are not the only ways in which galaxies interact. There may also be head-on collisions, disrupting galaxies and perhaps producing irregulars as fragments, and more gentle merging between spirals which also disrupt the disk systems but leave behind a different kind of system—an elliptical galaxy. These even trickier problems are only now being tackled by the computer simulations, but they show encouraging evidence that ellipticals really can be made in this way, from merging spirals. In systems clustered as closely together as galaxies are seen to be today, the computer simulations show that at least 20 percent of the original galaxies will be merged into ellipticals or distorted by the tidal effects of close encounters. Spirals don't just "lose" their disks to become ellipticals, but are reworked by merging, which completely changes the structure of the resulting galaxy. Instead of all the stars in an elliptical circulating the same way around the central nucleus (like the stars in a spiral galaxy), there should be two or more interacting families of stars, each of which follows an orbit which depends on which of the interacting galaxies it came from and the nature of the merging collision. And this is just what the observations show—Doppler evidence indicates that ellipticals are not rotating like single gas clouds, but contain stars which follow highly elliptical orbits, flavored by the velocities of the merging galaxies. In specialists' jargon, the velocity distribution of stars in ellipticals is highly anisotropic.

So it is no wonder that attempts to explain the nature of elliptical galaxies using "models" based on the collapse of gas clouds don't work—ellipticals simply did not form from collapsing gas clouds! The clarification that results from the new model resolves a lot of old problems, and it also means that a lot of interesting ideas previously put forward in an attempt to resolve some of those problems are no longer really tenable. It is a measure of just how much trouble theories of galaxy formation were in that some of these ideas ever got off the ground; my own favorite for a long time involved the idea that "white holes," cosmic fountains of matter best imagined as the opposite of black holes, might exist at the centers of all galaxies, spewing material out to become stars, planets, and so on. Alas for the romantics such as myself, there is no need to invoke such exotic extremes when tidal effects, merging, and isothermal fluctuations can now be seen as adequate to do the job. I still believe that white holes may play a part in the story of the Universe, perhaps tied to black holes through a kind of "cosmic subway" of wormholes through space-time.* But it is no longer necessary or reasonable to invoke them as the driving force of galaxies. It might seem surprising that all those Population III stars—90 percent or more of the matter in the Universe—should have left so little visible trace on the Universe. But their heyday was long ago—more than 90 percent of the history of the Universe has happened since the recombination era—and it may be that they have, after all, left a faint signature behind.

The excitement of the discovery of the cosmic black body background radiation in the 1960s gave way in the 1970s to ever more sophisticated techniques for measuring the exact spectrum of the radiation and determining, in particular, the extent of its very small deviations from a "perfect" black body spectrum. One of the most significant deviations is a blip in the measurements at wavelengths of a few millimeters or less. And, late in 1979, Michael Rowan-Robinson, of Queen Mary College in London, and col-

* See my book *Timewarps* (New York: Delacorte, 1979).

leagues working at the Berkeley campus of the University of California, showed that this blip might be a direct consequence of the first phase of star formation—the signature of Population III.

The blip occurs at just the right place to be the trace of the radiation emitted by hot silicate dust grains in space after the fireball era but still early in the history of the Universe. There is only one way in which such silicates could have formed so early, and that is if stars formed earlier still, burning their nuclear fuel to build heavy elements and blowing some, at least, of the material into space. Everything fits into the Population III picture, with one added bonus for those who, like myself, prefer the idea of a "closed" Universe to the prospect of infinite and eternal expansion. As Rowan-Robinson and his colleagues put it, "The energy input needed to distort the spectrum so drastically is substantial, requiring most of the matter in a Universe with the closure density to undergo thermonuclear reactions in Population III."*

In other words, unless there is enough matter in the Universe to make it closed rather than open, then it is hard to see how there could have been enough silicate produced by processing in Population III stars to produce the observed blip in the cosmic background spectrum. This is a very dramatic conclusion to be able to draw from the subtle ways in which the weak hiss of background radiation, only detected at all a decade and a half ago, deviates from a perfect black body curve, and it indicates the way in which subtle, second order effects of this kind can be used to provide as valuable information about the Universe as the insights provided by the headline-making discoveries. It does look as if we know where galaxies like our own come from, and the kind of Universe in which they were formed and live out their lives. The story of how a star like our Sun is born, along with a family of planets, and lives out its life in a galaxy like our own Milky Way system can therefore, for the first time ever, be set against this background of a consistent and self-contained story of how the whole Universe has got to the state it is in today from the Big Bang of creation. But our Galaxy isn't just a leftover flicker of the former glory of the Population III superhalo; it is also a relatively quiet backwater of the Uni-

---

* In their 1979 *Nature* paper (see Bibliography).

verse today, compared with the violent processes that do go on in some galaxies, and in the objects known as quasars. So before coming down to the homely problem of where our own Solar System comes from, this might be the appropriate place to pause for a slight diversion to look at the fierce beasts that lurk at the hearts of many galaxies. And it may not be such a detour after all, since it is just possible that a similar rough beast, now dormant or in hibernation, lurks at the heart of our own Milky Way.

VIOLENCE IN THE UNIVERSE

The activity of some galaxies shows that, while the Universe may be a quiet place today compared with events in the first thousand million years after the Big Bang, it is still capable of enormously energetic outbursts by any human scale. Some active galaxies look peculiar on optical photographs; some are detected by their strong radio emission. Many active galaxies are peculiar both at optical and radio wavelengths. The history of the study of peculiar galaxies goes back only to 1943, when Carl Seyfert published details of six galaxies that formed the archetypes for a class now called Seyferts in his honor. These are all spiral galaxies with small but very bright centers; about 1 percent of all spirals are now thought to be Seyferts, and this is best interpreted as meaning that *all* spirals spend about 1 percent of their lives in an active state of Seyfert behavior, rather than meaning that those one in a hundred observed in the Seyfert state today are particularly unusual.

Over the past forty years or so, many other kinds of active galaxies have been found, including the N-galaxies, which have even brighter nuclei than Seyferts but are otherwise very similar to them, and the quasars, which have nuclei so energetic that they completely outshine the rest of the galaxy in which they are embedded, and give the appearance of a single bright star, but at the distance of a galaxy. For some time after the discovery of quasars in the early 1960s there was a furious debate about whether the red shifts, which placed them at galactic distances, could be trusted as simple Hubble law indicators in their case, or whether they were a special case, quite nearby and showing high red shift for reasons unconnected with the expansion of the Universe. But improved observations over two decades showed that quasars really are embedded in the hearts of galaxies, and few astronomers now doubt the straightforward Hubble interpretation of their red shifts. This makes some of them the most distant objects known, a few with recession velocities greater than 90 percent of the speed of light. The standard view now is that Seyferts, N-galaxies, and quasars form a continuous progression in terms of energetic outbursts from the nuclei of galaxies, and that other peculiar extragalactic objects also fit into this scheme.*

All these objects are categorized by their optical appearance, and their peculiarities show up chiefly through the concentration of brightness (which means a concentrated region of energy emission) in the centers of galaxies. Radio observations, on the other hand, show how some energetic galaxies and quasars spread their influence across enormous stretches of space. A typical radio source will have two "lobes" of strong radio emission stretching out on either side of the galaxy to which it belongs. The structure is almost certainly the result of energetic particles being flung out in two opposite directions from a central compact source, and interacting with the tenuous gas of intergalactic space. The extended radio structure may extend over several million light-years, being energized by a compact region less than one light day across. And this, together with the rapid variability of some of these sources, suggests very strongly that they are driven by power from massive, central black holes. This in turn suggests that *all* galaxies may have black holes at their centers, since the differences between types of active galaxy seem to be differences of degree, not differences of kind, and the quietest phase, Seyfert activity, seems to be something that happens to all galaxies from time to time.

To put the energy of quasars in perspective, take the example of the source known as AO 0235 + 164, which flared up suddenly at both optical and radio wavelengths in the autumn of 1975. Over a few weeks, the power output from this quasar increased by more than 10^{41} watts, equivalent to 10,000 times the entire power out-

---

* For example, there is a class of objects rather like quasars called BL Lacaertae objects, which were at first regarded as some new kind of energetic galaxy but which look more and more like first cousins to quasars as more and more observations are made.

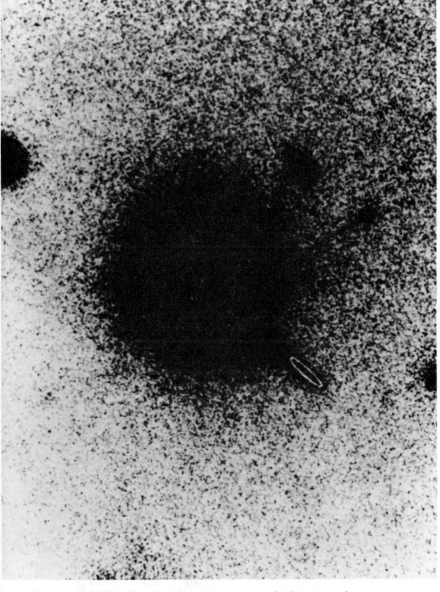

The quasar 3C 273. The white dot at the center marks the region of most intense
radio emission, and the white ellipse outlines a jet of material being blasted away
from the center and producing a secondary radio outburst. Almost certainly, this
kind of intense activity is associated with a black hole at the heart of the quasar.
(*Photograph from the Hale Observatories*)

put of our Galaxy at all wavelengths. Where could the energy come from?

In a hypothetical complete conversion of matter into energy (by matter/antimatter interaction) the conversion would be mc^2. But in the real Universe since the fireball era energy has only been available from less efficient conversion processes,* so that the mass equivalent of the energy liberated in such an outburst must be only a small fraction of the total mass present at the heart of the quasar. The mass involved runs into millions of times the mass of the Sun; yet the volume in which the activity occurs covers a diameter no bigger, in round terms, than the diameter of our own Solar System. So much matter in such a small volume must be in the form of a massive black hole.

The concept of black holes has seeped into general awareness, and is one of the few stock scientific jokes (along with the equation $E = mc^2$ and a couple of others) in the repertoire of newspaper cartoonists. Unfortunately, popular myths are not always entirely accurate sources of scientific information, so I should, perhaps, set the record straight on just what a black hole is.

BLACK HOLES AND QUASARS

In a nutshell, a black hole forms when enough matter is concentrated in a small volume so that the inward pull of gravity is strong enough to overcome any resistance and the matter collapses into a state so dense and with such a strong gravitational field that not even light can escape from it. Any collection of matter—the Earth included—has a gravitational field at the surface whose strength depends on both the mass and the radius of the object. To escape from a planet such as the Earth, a rocket or other projectile must be given enough energy to overcome this gravitational attraction. An object thrown off with less than the critical "escape velocity" will fall back, but one thrown off with more than escape velocity escapes into space. For Earth the escape velocity is 11.2 km per second; for the Moon it is 2.4 km per second. The surface

---

* And, of course, since AO 0235 + 164 is still there, we know it wasn't converted into pure energy in the outburst!

escape velocity is bigger if the mass of material inside is bigger, so that a planet like the Earth in size and shape but twice as massive would be much harder to escape from. But since giant planets like Jupiter are not just more massive than Earth but bigger as well, the escape velocity from the outer fringes of their atmospheres may not be much larger than that from Earth. The two factors, size and mass, must be balanced against one another in determining escape velocity, so that even for our Sun the surface escape velocity is only 617.7 km per sec, while for Jupiter it is 61 km per sec.

The Earth is held up against the inward pull of gravity by the electric forces between atoms, which prevent them from merging into one another. Inside the Sun, where electrons have been stripped from the atomic nuclei, the heat from the fusion processes going on in the interior keeps the nuclei in motion like atoms in a hot gas, producing an outward pressure which holds the Sun up against the pull of gravity. And when the nuclear fusion processes end, the Sun too will collapse down into a quiet, compact state where gravity is opposed by the electric forces keeping atoms apart. But for a star only two or three times as massive as the Sun, once the center cools and the outward pressure drops, gravity is com-pletely dominant. Atoms are literally crushed out of existence by the irresistible inward force, and collapse continues indefinitely. As this happens, the density increases and the radius of the star shrinks, producing a huge value of the escape velocity. There is no longer any "surface" to use as a reference, except for the boundary radius at which the escape velocity from the shrinking star exceeds the speed of light. Once the object has collapsed inside this radius, no information from the inside can ever get out, as light itself is trapped by the intense gravitational field. For all practical pur-poses, the star has disappeared, and the surrounding sphere where the escape velocity is the speed of light is the surface of a black hole.

One way to make a black hole, then, is to put a mass of more than a few Suns' worth of material in one place, wait for nuclear fusion reactions to run their course, and watch it collapse.* It is also possible to imagine making a black hole another way. Because

---

* This, however, takes a long time—several thousand million years!

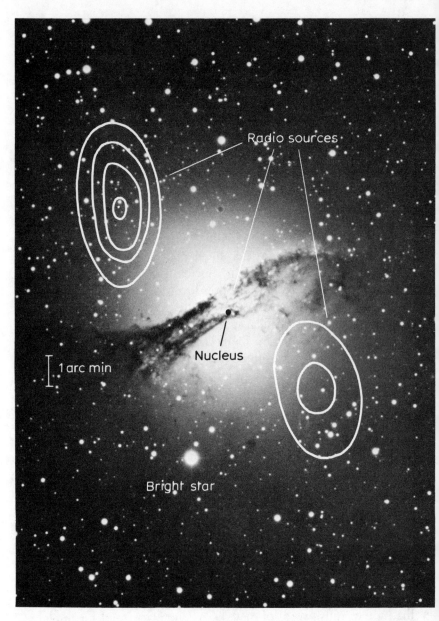

Radio sources

Nucleus

1 arc min

Bright star

The galaxy Centaurus A may also harbor an energetic black hole. As well as the peculiar structure of the galaxy itself, there are radio sources both at the center and on either side, suggesting energetic outbursts sometime in the past. The bright object just below the galaxy is a foreground star in our own Galaxy. (*Photograph from the Anglo Australian Telescope*) the escape velocity at the surface of an object depends inversely on the square of the distance from the center but directly upon the mass, which itself depends on the volume and is proportional to the cube of the radius, the net effect is that if more mass with the *same* density is added to an increasing volume, then the escape velocity at the surface increases. So if we imagine a cluster of stars like the Sun stacked side by side in space, the escape velocity from the surface of the cluster is bigger than the escape velocity from the surface of the Sun. For a big enough cluster, the escape velocity is bigger than the speed of light, and it becomes a black hole from which nothing can escape. This is very curious, since it would be quite possible for a whole galaxy to be compact enough (that is, dense enough) for this to happen, without conditions inside being very different from those we see around us in our Galaxy—stars could form, with planets and life on them, and all within a low-density, large-radius black hole. Indeed, this is one way of looking at the Universe itself, if it is expanding at less than the critical velocity needed for it to be "closed." Such a Universe is, in a real sense, a self-contained black hole. But these are very different objects from the beasts that reside at the centers of galaxies.

Those really must be the superdense variety of black hole, with matter crushed out of existence behind the barrier of the surface at which the escape velocity exceeds the speed of light, the "horizon" of the black hole. It is not clear what happens to the matter beyond the horizon; perhaps it disappears altogether from our Universe. (Perhaps it even reappears as a "white hole" in someone else's universe!) But what matters, in terms of the activity at the centers of galaxies, is how a black hole behaves and how it looks from the outside. In a nutshell, it is the ultimate bottomless pit, swallowing up any matter that comes within its gravitational clutches and growing more massive, with a stronger gravitational field still, as it does so. As matter is ripped apart and sucked into the hole, particles are constantly in collision with one another in a swirling vortex of material trying to funnel into a tiny volume of space. And with all this jostling the particles get hot and radiate energy across the electromagnetic spectrum. Under the right circumstances, the energy conversion process can be more efficient than anything except matter-antimatter annihilation, with 10 or 20 percent of the rest mass of the infalling material being converted into energetic radiation. And that brings us right back to outbursts like the autumn 1975 flare-up of AO 0235 + 164.

For an efficiency of mass-energy conversion of 10 to 20 percent, a typical quasar must be gobbling up between one and two solar masses a year—a very reasonable amount of matter by the standards of a surrounding galaxy which might contain several thousand million stars, suggesting a fuel supply that could last for millions of years, building up a central black hole of perhaps a hundred million solar masses and still leaving plenty of stars in the surrounding disk to settle down as an ordinary galaxy after the quasar activity had died away. The collapse of a trace of residual gas inward to the center of an old superhalo provides several ways in which the original central black hole might form—the heart of the gas cloud itself might collapse right through to the black hole state right away, or it might produce a few very massive stars which explode, leaving black hole remnants which merge with one another at the heart of the embryo galaxy. It is even possible that a central cluster of a hundred million stars could form, run through their life cycles to produce a cluster of black holes, and leave these holes to merge into one, all during the early stages of the formation of the galaxies as we know them today. After all, the phase of fireball activity and the formation of the original superhalos happened within about the first 2,000 million years of the 15,000 million-year history of the Universe; the collapse of residual gas inward to the halo centers should produce a burst of violent activity until things settle down into a new orderly pattern (typically a spiral galaxy), still leaving perhaps 10,000 million years before we get to the present state of, say, our own Galaxy. And all of this explains, rather neatly, why the brightest quasars are at the biggest red shifts. Remember that the further away an object is the longer its light has taken to reach us; for a quasar with a high red shift we may be seeing it by light which left when the Universe was only 4,000 million years old and the black hole was still absorbing surrounding matter. But when we view a galaxy with a red shift corresponding to a much smaller recession velocity, we may see it as it was 4,000 million years ago, when things had settled down a lot more.

Although the greatest burst of massive black hole formation and quasar activity should have occurred early in the history of the Universe—which is borne out by the fact that we see more quasars at high red shift than nearby—the process could still be continuing in some galaxies that have evolved more slowly, explaining the quasars that are seen relatively nearby. And the intermediate stages of the process, the various types of black hole precursors which might form at the centers of galaxies, could well explain the variety of active galaxies intermediate between "normal" galaxies and quasars, such as N-galaxies, Seyferts, and BL Lacaertae objects. Alternatively, Seyfert outbursts may occur when the old central black hole formed in the original quasar activity manages to capture some material from its surrounding galaxy, briefly flaring up in a small-scale return to its original days of glory. This is the explanation which I find most satisfactory as an explanation of why every spiral spends about 1 percent of its life in intermittent bursts of Seyfert activity. The material could be supplied to the central black hole by gas from stars, the debris of stars which have collided with one another near the central nucleus, or even whole stars ripped apart by the tidal forces of the black hole as their orbits take them dangerously close. For, remember, a quasar outburst needs only one or two solar masses per year to power it, and a Seyfert galaxy is only a fraction as powerful an energy producer as a quasar.

Any really satisfactory model of quasars must explain how the "double lobe" radio patterns form, and the black hole theory does just that. The black hole must be rotating, and quite rapidly considering how much the original gas cloud has contracted in forming it, so that the swirling material around it is concentrated in a disk. In such a situation, as energetic particles are produced in the collisions between atoms funneling into the hole, they will escape along the "lines of least resistance" from the "poles" of the spinning system. The biggest known radio source extends over 20 million light-years, showing that a steady production of energetic particles has gone on for 20 million years, with the beams directed along the same precise alignment for all that time. In addition, there is a very compact central source in the middle of the galaxy (3C236) which is also aligned in precisely the same direction. This pattern exactly fits the idea of a rotating black hole blasting out particles from its poles.

And what of the very powerful, rapidly fluctuating sources like AO 0235 + 164? Almost certainly, these are cases where the jet just happens to be pointing straight toward us—we are, in effect, looking right down on the massive, rotating black hole from above.

We know, from the size of sources like 3C236, that quasar activity can continue for about 20 million years. But there are no bigger sources, which hints that the activity cannot continue for much longer than this and backs up the observation that there are few strong radio sources and active quasars left here and now in the Universe. There must be many dead quasars lying at the hearts of galaxies, however, and it is interesting to look at our near neighbors in the Universe to find evidence of their presence. The best candidate for a dead quasar is probably the radio galaxy Centaurus A, just about five Megaparsecs away. This has a large pair of radio lobes which now radiate only weakly compared with objects like 3C236 or AO 0235 + 164, plus a tiny radio and X-ray source at the heart of the galaxy only one light day across. This is probably a black hole of about 10 million solar masses, shining by the energy released as remnants of material trickle into it.

But the most interesting galaxy from a human point of view is the one we live in. Is there any evidence that our Milky Way Galaxy was once a quasar? There is certainly strong evidence of violent activity at the heart of the Milky Way. Radio observations reveal clouds of gas moving outward from the center, and even thin jets of material moving away from the nucleus. At the heart of our Galaxy there is a rotating disk of material some 1,500 parsecs across, with an expanding ring of cool gas, 380 pc across, embedded in it. The central region contains clouds rich in every kind of molecule detected in interstellar space, and there is a strong (by galactic standards) radio source right at the center of the Galaxy. This source is only about a thousand million kilometers across, and could very well be the result of a slow trickle of accretion onto a central black hole. But some idea of the mass of the object at the nucleus can be worked out from our old friend the Doppler shift, which reveals the speed with which the central clouds are orbiting the center. A more massive central object would keep the clouds orbiting at higher velocity (in just the same way that a more massive object has a higher escape velocity), and the observed velocity pattern suggests a central object of about 5 million times the mass of the Sun. This is not enough to power a full-blown quasar— although it is impressive for an object small enough to fit within the orbit of Jupiter around our own Sun. But it could well be enough to power Seyfert outbursts when large enough concentrations of matter fall into it.

From the distribution of the major features which are now detected moving rapidly away from the center, we can even make a good guess as to the time the last major outburst occurred in the nucleus of our own Galaxy. This comes out as only 12 million years ago, a tiny fraction of the history of the Galaxy, or of our own Solar System, and hints that in the very recent past, by cosmic standards of time, our own Galaxy was in the Seyfert state. A few astronomers have speculated that this kind of outburst, occurring every couple of hundred million years or so, could send its influence rippling across the Galaxy to disturb our own Solar System and even Planet Earth itself, producing stresses (perhaps in the form of floods of cosmic radiation) which could explain the rather sudden "extinctions" of many species of plants and animals which are found in the geological record. It is also possible that the activity may account for the spiral structure seen in the disk of our own and other galaxies. In simple calculations of how spiral patterns should change as a galaxy rotates, it turns out that the pattern should last for no more than about a thousand million years before being smoothed away by rotation. The apparent permanence of the pattern in spiral galaxies can be explained by a density wave theory, in which the spiral pattern is thought to be a standing shock wave moving around the disk, and if the central black hole is involved in outbursts every few hundred million years, then each such upheaval could generate its own spiral pattern, a result of a combination of an outward blast and the roughly circular motion of stars in orbit around the central nucleus. In that case, the pattern would be repeatedly renewed, and each particular pattern might well last for much less than a thousand million years between upheavals. The shock—or blast—wave theory is fine provided you can explain where the blast comes from, and the idea of an active galactic nucleus, a 5 million–solar mass black hole, involved in repeated Seyfert outbursts seems to fit the bill nicely.

It would be surprising if all of this activity had no effect on the Solar System, and, indeed, the best current theories see a very close relationship between the formation and evolution of the Solar System and the nature of the spiral Galaxy in which we live. The spiral structure, in turn, now seems to be related to the nature of the central black hole, and this and the bright Galaxy itself are just leftovers from the formation of a huge supergalaxy out of an isothermal fluctuation in the recombination era 19 thousand million years or more ago. So by the time we start to look at just how the Sun and its family of planets formed, we are already a long way down a chain of interlinked and, some might say, improbable events. And we haven't yet even begun to consider the puzzle of the origin of life!

3 THE ORIGIN OF OUR SOLAR SYSTEM

The origin of our Solar System is intimately connected with the nature of our Galaxy, just as the origin of our Galaxy resulted from the underlying structure of the whole Universe. At every step down the chain from Big Bang to man himself we will find this close relationship with what has gone before; and ultimately this tells us that the kind of creatures we are depends not just on the kind of planet we live on or the Sun that is in our sky, but on the detailed physical nature of the Big Bang—the origin of the Universe—itself. We are creatures of our Universe, and creatures like us could not have evolved without the Universe being what it is.

When we look at the problem of how our own Solar System— the Sun with its attendant family of planets and cosmic debris— came into being, we are, however, starting to ask much more specific questions than those we have addressed so far on the nature of the Universe and the origin of galaxies. Across that broad sweep of space and time, the answers must inevitably be quite general, and therefore "easy" in the sense that they do not deal with specifics. It is a slightly different kind of problem, though, if we wish to explain why a star just the mass of our Sun should exist, with a family of just so many planets, each at particular distances from the Sun, and each with some specific mass. Now, because we are asking more precise questions, the "answers" theorists can come up with look, in many cases, more vague. But we can still describe in general terms how a Population I (disk population) star is born out of a cloud of interstellar gas, with an attendant retinue of planets.

Since our own Sun is a member of the disk population of our own spiral galaxy, and since my focus of attention in this book is on how *we* have been produced out of the chaotic fireball in which the Universe began, at this stage in the story we have to deliberately pass by some intriguing puzzles and features of not just the Universe at large but of our Galaxy at large. Since our own Galaxy is not a quasar, or a strong radio source, or an elliptical, for example, the story of the development of human life within our Galaxy does not depend on the nature of quasars, strong radio galaxies, and ellipticals; since our Sun is a member of the Population I disk, the story of the Solar System does not depend in any detailed way upon the nature of the older stars of the Population II halo, and still less on the stars of the Population III superhalo, once we have acknowledged that the presence of the disk, and its rotation rate, was determined long ago when the proto-Galaxy was first forming from a fluctuation in the dying fireball. The central question, in human terms, is how Population I stars, and their planets, form within the rotating disk that is characteristic of a spiral galaxy like our Milky Way system.

A GALAXY OF STARS

Seen from above, the disk of our Galaxy would look like a huge catherine wheel of stars, with a central bulge some seven kiloparsecs across surrounded by tightly wrapped spiral arms covering a total diameter of some thirty kiloparsecs. The Sun lies about two-thirds of the way out from the center to the rim, about 10 kpc from the nucleus of our Galaxy. This is a strikingly unspecial place—our Solar System is neither close to the nucleus nor at the very rim of our Galaxy, but seems to be a very ordinary system in a very ordinary part of the Galaxy. At the distance of the Sun from the center, the disk is about 800 parsecs thick; the nuclear bulge is about 3 kpc thick, and further out than our Solar System the disk thins a little.

The whole system contains about 100,000 million stars, which is on the large side for spirals—not spectacularly so, but enough to justify the classification of our Milky Way system as a "giant spiral." So at least there is *something* slightly out of the ordinary about our home in the Universe!

All of this information comes, of course, not from some imaginary observer looking down on the disk, but from painstaking observations from the Solar System's position embedded in the disk two-thirds of the way out to the edge. It is rather as if an observer were set down in a wood with a compass and some kind of rangefinder, so that without moving he could plot the positions of all the visible trees and draw a map of his part of the wood. If the trees were distributed in some regular pattern, he could then fill in the blanks on the map—the trees out of sight—by guessing where trees ought to go to continue the regular pattern. In our Galaxy, the regularity is provided by the spiral structure; optical, infrared, and radio observations (especially at the 21-cm wavelength characteristic of radiation from hydrogen clouds) provide the observer's eyes; and range finding depends on a variety of techniques, but especially the regular behavior of Cepheid variables mentioned before. With velocities determined from the ubiquitous Doppler shift (surely the single most valuable "tool" of observational astronomy), we know about the rotation of the whole Galaxy, as well as the motion of individual stars and clusters of stars.

It is the rotation which holds the disk "up" against gravity, of course, but only in the plane of the disk. In a sense, the disk itself is still forming—certainly still evolving—since new stars are still being born in it. And, as the material in the Galaxy (strictly speaking, the leftover material from the long gone glory days of the superhalo) is still settling down under the pull of gravity and through successive collisions between gas clouds, the region of activity in the disk is getting thinner. The oldest disk stars are spread over a total thickness of 700 to 800 parsecs; the youngest, newly formed stars (including many massive, hot O stars) are confined within a narrow region only 80 pc thick. And our Sun, once again showing its unexceptional mediocrity, is a middle-aged disk star which wanders no more than 80 pc above or below the plane of the disk.

The stars of the disk form in groups, and the stars themselves tend to be small, most being only about one-tenth of the mass of our Sun. So at least the Sun is bigger than average, although by no means a giant in stellar terms. It seems that disk stars form in loose clusters, with many stars being born out of one collapsing gas cloud, and that the clusters are gradually dispersed as the individual stars orbit around the Galaxy. The ages of clusters can be determined from the colors of the stars, which depend on their surface temperatures and change as each star ages; some clusters are known which are a few million years old, and others have ages ranging up to a few thousand million years. Altogether, more than 700 of these "open" clusters have been identified in the immediate vicinity (within 2 kpc) of our Solar System. What makes a gas cloud collapse to form a loose cluster of stars? That, it seems, depends very much on the structure of the disk itself.

First, the disk does not rotate like a solid wheel. Each star—and each solar system—orbits around the center of the Milky Way at its own speed, which depends on the distance it is from the center. Close in, stars have to orbit quickly to avoid being pulled into the center by gravity—so only fast moving stars are left there, the rest having been gobbled up by the central black hole. Farther out, where the gravitational pull is weaker, a more leisurely pace will hold a star secure in its own roughly circular orbit. So the disk as a whole is in differential rotation, with its inner regions spinning more rapidly than the parts farther out. But such terms as "leisurely" are relative—in the case of our own Solar System, the orbital velocity is 250 km per sec (that is, we are moving through space around the Galaxy at 900,000 km per hour!), which seems impressive enough. However, even at that speed it takes the Sun and Solar System roughly 250 million years to complete one circuit of the Galaxy. Since the Solar System formed, it has circumnavigated the galactic disk only twenty times, in about 5,000 million years. Even so, the fact that all the stars in the disk are in constant differential motion means that the spiral pattern that is so distinctive now ought to be broken up and "dissolved" in a short time—certainly much less than the 250,000 years it takes the Sun to orbit the Galaxy. The standard explanation of how the spiral pattern persists even while the stars follow their own independent orbits is that the true spiral features are not the bright stars which stand out so well in astronomical photographs, but the dark lanes of gas and dust which lie alongside them. The spiral arms are essentially regions where gas is concentrated and compressed, and bright stars are actually born along their edges as a result of the compression. This explains the pattern of bright stars, but it does not entirely explain the spiral structure since there is no established theory of why the gas and dust should form a spiral pattern in the first place! Almost certainly this is connected with violent activity in the nucleus of the Galaxy, where shock waves moving outward from the center interact with gas moving in circular orbits to set up a spiral pattern. What matters for the formation of the Solar System, however, is that a spiral pattern of concentrated material (a "density wave") did exist from early in the history of the Milky Way's disk system, and that this density wave provided a repeated squeeze for a certain gas cloud which was orbiting around the center of the Galaxy at a distance of some 10 kpc.

BIRTHPLACES OF STARS

Each time such a cloud crosses a spiral arm (which happens twice in each orbit, once on either side of the Galaxy, if there are two arms as in most spirals), it is squeezed by the compression wave. It may "unsqueeze" a little when it comes out on the other side, but it should still end up smaller and denser (which means more tightly held together by its own gravity) than before the passage. And after a few orbits around the Galaxy, being squeezed a couple of times in each orbit, such a cloud will collapse completely and fragment into protostars, as the strength of its own gravitational pull becomes enough to break the cloud into pieces. In some cases, the "last straw" that triggers the final collapse of the cloud may be an extra shock wave, blasting outward from a nearby supernova explosion in which a massive star has violently come to the end of its life. These massive stars themselves form in the dense material of the spiral arms, but may live for only a few million years, exploding before they have moved far off around the Galaxy. And the shock waves they produce can set off star formation in other clouds in that part of the spiral arm. This, in fact, is what seems to have happened at the birth of our own Solar System.

Supernova explosions mark one possible way for a star to die, and I shall explain how some stars get into an explosive state in due course. For now, all we need to know is the fact that some stars do explode, shining for a few days as bright as a whole galaxy of normal stars, and sending shock waves ripping across the tenuous gas and dust clouds of interstellar space. They also spread material from the exploding star through space, mixing this in with the material of the interstellar clouds and enriching the mixture from which new stars (and planets) will form. It is this supernova material which provides strong evidence that our own Solar System was born in the aftermath of such a stellar explosion.

The evidence comes directly from meteorites, fragments of interplanetary rock that fall to Earth from time to time. These are thought to be leftover bits of the material from which the Solar System formed, still orbiting around the Sun so that every now and then some of them collide with the Earth. The smallest fragments burn up in the atmosphere as meteors, and this meteoric bombardment continues incessantly as sand-grain–sized particles rain down on the Earth. Chunks big enough to survive the fiery fall through the atmosphere reach us less often, and most of those are never found since they impact over the sea or in thinly populated desert, jungle, or mountain regions. Some enormous meteorites have hit the Earth during its lifetime, producing craters like those of the Moon, most of which are now washed away by erosion but a few of which, notably the great "Meteor Crater" in Arizona, are still visible. And a few modest-sized chunks of meteorite have been found on Earth and studied in scientific laboratories.

These studies received a great boost in the late 1960s, when improved techniques were developed primarily to help in analyzing the Moon rocks brought back to Earth by Apollo astronauts. In 1969 a large meteorite fell near the village of Pueblito de Allende

A glowing cloud of gas and dust in space marks a site of active star formation in the Large Magellanic Cloud, a small irregular galaxy close to our Galaxy. The cloud shines because the hydrogen in it is energized by the hot young stars inside, in the same way that gas in the tubes of advertising signs is energized by electricity. Soon—on the galactic timescale—the hydrogen will blow away into space, leaving the new stars behind. (Photograph of 30 Dor by the Anglo Australian Telescope) *Credit: Science Research Council*

in Mexico, with some two tons of material surviving the fall and fragmenting into recoverable pieces. This Allende meteorite has since been analyzed in more detail than probably any previous meteorite samples, in the belief that it contains materials unchanged since the time of the formation of the Solar System, nearly 5,000 million years ago.* That material differs in some important ways from the material of the surface of the Earth, which has, of course, undergone some drastic changes since the Solar System formed. The key evidence comes from the amounts of certain isotopes of particular elements in the meteorite samples.

Elements come in different varieties, called isotopes, which are chemically identical but have atoms of different weight. Oxygen, for example, is most commonly found to be the isotope oxygen-16, each atom of which is made up of a cluster of eight positively charged protons and eight neutral neutrons (the nucleus) surrounded by a cloud of eight electrically negative electrons. Two other isotopes are oxygen-17 and oxygen-18, which differ from oxygen-16 only in having one and two extra neutrons, respectively, in their nuclei. This does not affect their chemical behavior, which depends on the number of electrons and their arrangement; but it does make them physically distinguishable since oxygen-17 is heavier than oxygen-16, and oxygen-18 heavier still. All of these oxygen isotopes are stable. But some isotopes of some elements are radioactive, spontaneously emitting particles and being transformed into a different element. Aluminum-26, for example, contains thirteen protons and thirteen neutrons in its nucleus, but is unstable (the stable form, aluminum-27, has one extra neutron). The instability shows itself because atoms of aluminum-26 will each, in time, emit one positron (a positively charged counterpart to an electron), converting a proton in the nucleus into a neu-

* How do we know the age of the Solar System? The best "clock" is provided by the radioactive decay of materials such as uranium-235 and uranium-238. Over millions of years, these radioactive atoms break up, eventually producing stable atoms of lead, and along the way giving off helium which can be trapped inside the frozen material of a meteorite in space. When the meteorite hits the Earth and is recovered, it is possible to measure the amount of helium and compare this with the amount of uranium still left in the material. The older the sample is, the more helium there will be in proportion to uranium; and although this sounds pretty bizarre as a clock mechanism, it is, in fact, accurate enough to set the age of the oldest meteorites (and, therefore, of the Solar System) as a little over 4,500 million years and certainly less than 5,000 million years.

When a star dies in a supernova explosion, it scatters material across space in great streamers of gas. Eventually, this material, including heavy elements, becomes part of clouds of gas and dust in space that collapse to form new stars. (*Lick Observatory Photograph*)

tron and changing the atom into magnesium-26 (twelve protons, fourteen neutrons). The very regular behavior of radioactive decay processes means that for any chosen sample of a particular radioactive isotope, half of the atoms will decay in this way in a time characteristic of that particular isotope. In the case of aluminum-26, the "half-life" is 700,000 years, so that any Al-26 present at the formation of the Solar System must itself have been formed only shortly before, on the cosmic timescale, and will all have decayed by now, leaving magnesium-26 behind.

This is just what we find in the Allende meteorite samples—magnesium-26 which could only have been formed by the decay of aluminum-26 inside the meteorite after the Solar System formed. (The two other stable isotopes of magnesium, magnesium-24 and -25, could form in other ways, but it happens that isotope 26 is *only* made from the decay of aluminum-26). So the cloud from which the Solar System collapsed contained aluminum-26, which must have been produced somewhere and scattered across the cloud just before it collapsed.*

In addition, the proportions of the oxygen isotopes found in the Allende meteorite samples are different from that on Earth—there seems to be extra oxygen-16 present, and this too must have come from "outside" the collapsing gas cloud. And there are further subtle differences between the isotope abundances in the meteorite samples and those which we would expect if the cloud had simply collapsed quietly under its own gravitational pull. The best way to make exotic isotopes is in a great supernova explosion, a process so energetic that atoms are ripped apart and atomic nuclei are put together into new and exotic arrangements before being hurled across space. The only way in which the isotope abundances found in meteorites can be explained is if a supernova exploded alongside the gas cloud which gave birth to the Solar System just before it collapsed. In that case, it seems more than likely that it was this supernova explosion that triggered the collapse—and, using yet more subtle measurements of isotope abundances, astronomers such as David Clark, of the Royal Greenwich Observatory, estimate that a previous supernova had occurred near the

---

* "Just before" here means within a million years or so, an eyeblink on the scale of the time it takes the Sun to travel once around the Galaxy.

The best modern telescopes and latest photographic techniques bring out a spectacular wealth of detail, as in this photograph of a streamer of interstellar material, the stuff of which new stars and solar systems are made. (*UK Schmidt Telescope photograph, supplied by the Royal Observatory, Edinburgh*)

Messier 8, the Lagoon Nebula shows up several of the dark features known as Bok globules. These are patches of light-absorbing, dusty material in space. The nebula itself is 30 light-years across and 4000 light-years away.

(Photograph from the UK Schmidt Telescope, supplied by Royal Observatory, Edinburgh)

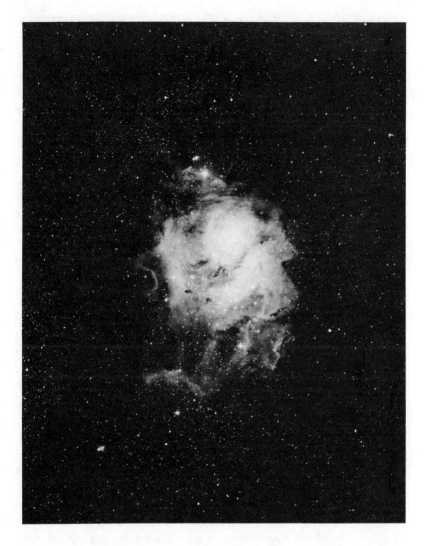

The Horsehead Nebula is a dramatic patch of cool gas silhouetted against softly glowing, hotter interstellar material. It is part of the complex of stars and nebulas near Orion's belt and sword; the bright blue star to the left of the "horse head" is the easternmost star in Orion's belt, Zeta Orionis.

(Photograph from the UK Schmidt Telescope, supplied by Royal Observatory, Edinburgh)

cloud about 100 million years earlier. This, it happens, is just the time it takes our Solar System to cross from one spiral arm to the next in its orbit around the Galaxy. So perhaps the first supernova helped to set the scene by squeezing the presolar cloud in one arm, then 100 million years later in the other spiral arm, the final squeeze came in the form of another supernova. Only then, at long last, could the gravity of the cloud take over. As the cloud broke up under the pull of its own gravity, it would have formed many stars which have since become separated from one another and scattered around the Sun's orbit through space. Just one fragment of that original cloud, dominated now by its own gravitational pull as its density increased, was destined to form the star we call the Sun and a family of planets, the Earth among them, which together make up our Solar System.

A WHIRLING DISK

The nature of the Solar System, as it condensed out of a collapsing cloud of gas, was determined by rotation. Just as a collapsing cloud of gas, on a much larger scale, was held up by its rotation into a disk which became the flattened Milky Way system in which we live, so the rotation of the presolar gas cloud held up some of the collapsing material into a disk. A great deal of material did indeed settle into a more or less spherical ball at the center—the Sun itself—but this could only collapse to such an extent because the angular momentum of the original gas cloud was spread over the remaining material in the disk. The Sun got most of the mass, but the disk got most of the angular momentum.

It might seem an exaggeration to describe the Sun's family of planets as a disk, but in terms of the Solar System as a whole there is no doubt about the validity of the description. It's not as if we just had one small body orbiting around another, like the Moon around the Earth. In the Solar System, there are nine known planets, all of which move around the Sun in the same way; most of these planets have moons, and the overwhelming majority of these orbit their planets in the same sense that the planets orbit the Sun, while the planets themselves, with the exception of Venus and Uranus, turn on their axes in the same sense again. This

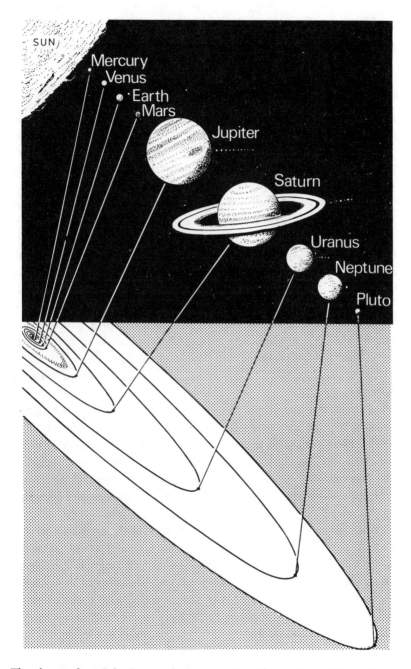

The planets of our Solar System, shown approximately to scale, and the positions of their orbits in the disk of material surrounding our Sun.

dominant rotation sense of all the material in the disk is also the direction of rotation of the Sun itself, which turns on its axis once every 25.3 days. And, finally, all of this material rotating the same way about the Sun is indeed concentrated in a thin disk, the plane of the ecliptic. This evidence very clearly shows the intimate relationship between the Sun and planets, and leaves no room to doubt that they formed together. If the planets had somehow been picked up by the Sun on its travels around the Galaxy, they would not be so regularly arranged, all in the same plane and with essentially circular orbits; if the Sun had collapsed on its own from a great gas cloud, it would be spinning much more rapidly than we see today—indeed, it could never collapse into such a compact ball as it is today. A collapsing gas cloud *has* to lose angular momentum in order to collapse enough to become a star, and the best way to do the trick is by growing a rotating disk of material. The material in the disk can then collapse into planets, with the biggest planets (especially Jupiter and Saturn) going through the same process in miniature, growing their own disks of material as they contract, some of which have formed moons and some of which remain as rings today, most notably the beautiful rings of Saturn.

Astronomers still debate the details of how planets formed out of the disk around the Sun. In a major academic work published in 1978* twenty-eight of the most eminent theorists of modern times spent no less than 668 pages covering the best modern thought on the problem of the origin of the Solar System, and still came out with no single "answer"—there is no one detailed model accepted by everybody. But a great deal of the disagreement between the theories *is* over detail, and does not represent any fundamental dispute over the evidence that Sun and planets formed together, going through a Sun plus disk stage as the original gas cloud collapsed. So although the following simplification of the current thought on the problem cannot be presented as correct in every detail, it does give a broad outline of the ideas that are becoming increasingly accepted today, and within which the details, al-

* S. F. Dermott, ed., *The Origin of the Solar System* (New York: Wiley, 1978).

though still being debated, are likely to be filled in in the not too distant future.

From this point of view there is one basic feature of the Solar System—apart from its existence!—that is of crucial importance to human life and must be explained by any satisfactory theory. Why is the Solar System made up of two kinds of planets—small, rocky planets near the Sun (including the Earth) and large, gaseous planets further out?* This can certainly be explained satisfactorily in terms of the evolution of a disk of material around the Sun, heated by the young star at its heart. The heating is of key importance, driving material out and away from the nebula in which the Sun formed. In order to explain the slow spin of the Sun, most modern theories require not just the transfer of angular momentum to the disk, but a lot of material—as much as remains in the Sun itself—blown out of the system altogether by the fiery furnace at its heart. Long before this furnace switched on, however, the processes that led to the formation of the planets had already begun.

Tiny grains of dust which existed in the original gas cloud (nebula) would begin to stick together to form little fluffy "supergrains," perhaps a few millimeters across, as the nebula started to contract and the grains collided with one another more and more often in the increasingly dense cloud. Constantly bombarded by the atoms of gas in the collapsing cloud, these supergrains would be very susceptible to the processes which transfer angular momentum, and would settle rapidly into the disk as it began to grow around the bulging, flattened ball of the embryonic Sun. This concentrated the supergrains into a still smaller volume of space, giving them ample opportunity to come into contact with one another and to interact through gravity. The result was that they clumped together into bigger lumps, building up to pieces a kilo-

---

* Pluto, generally regarded as the farthest known planet, is an exception to this pattern, being both far out and small and rocky. But Pluto doesn't seem to be a "proper" planet at all, since it follows an elongated orbit which brings it closer to the Sun than Neptune for some of the time, including from January 1979 to March 1999. In all probability, Pluto is a moon that has escaped from one of the gas giants, so that the "real" planets divide rather neatly into two groups of four, separated by a belt of cosmic rubble between Mars and Jupiter, the asteroid belt.

meter or more across, rather like meteors or the rubble that remains in the asteroid belt today, and forming the first respectably sized solid objects in the Solar System. As this process of clumping together (accretion) continued, eventually planet-sized objects were built up—although the details of the accretion process and the reasons for the exact positions of the planets from the Sun are among the puzzles still being debated by the experts.

TWO KINDS OF PLANET

It is straightforward enough to account for the two kinds of planet by this process, although, again, the details of the explanation are not fully worked out. Close to the Sun, the heat from the young star at the center was enough to drive out any material that is easily volatilized, so the growing clumps of matter were dominated by substances that are not easily vaporized, such as iron and silicates. Farther out from the Sun, however, the original grains of the nebula would retain a coating of water ice, frozen methane, and solid ammonia, too far away from the Sun for these materials to evaporate. And in the remaining gases of the nebula itself, surrounding the growing solid lumps, very light gases such as hydrogen and helium would get blown away and could only be retained by planets well away from the heat of the Sun. So it is entirely natural that small, dense, rocky planets formed near the Sun and that large gaseous planets formed in the outer Solar System. The process even explains some of the detailed differences between the four inner planets, since Mercury seems to be very dense and compact, almost like the Earth with its rocky outer layers stripped away, while the Earth and Venus are rather less dense, and Mars is less dense still, deficient (compared with Earth) in the heavy metals that make up much of the core of our own planet.

Even when the planets and their moons had formed, the young Solar System was still an active place. A great deal of gas and dust remained around the Sun, and there were still plenty of rocky asteroids being swept up in collisions, producing the scars which crater the Moon, Mars, Mercury, and Venus, and which are present, softened by erosion, on the surface of the Earth. The young Sun became hot, in the first place, simply because of the gravitational potential energy released as it contracted. It became a true star only when the center became hot enough to start up nuclear fusion reactions, which have kept it shining for the past 4 to 5,000 million years. And when that happened, before it settled down into a phase of steady nuclear burning, it must have produced an initial outburst of energy—heat—which blew away all of the remaining gaseous material that was not securely held to a planet by gravity.

All of this was a natural consequence of the collapse of a cloud of gas in space, with a smattering of a small percentage of dust in the form of interstellar grains. The tiny amount of dust became planets in orbit around a star, and there is no reason to doubt that every time a star forms in this way a planetary system is also produced. Very many stars—perhaps the majority—seem to be in multiple systems with two or more stars orbiting one another. That kind of system might not be suitable for the formation of planetary systems like our own Solar System. But wherever a star does form in isolation—and even if that is a minority of the disk stars, that still means millions of stars in our Galaxy alone—planets like those in our Solar System should form as well. There must be very many planetary systems in our Galaxy, and there may well be planets rather like the Earth in some of those systems. Does that mean that life—even intelligent life—may be common? That depends not just on the presence of planets, but also on the nature of the star at the heart of the planetary system. Some stars are more hospitable, in human terms, than others; and although from here on the story of where we come from must inevitably focus on just one planet within one Solar System, we can retain some idea of our place in the Universe by a quick look at how stars behave.

A VARIETY OF STARS

It is, of course, impossible for astronomers to sit and watch the evolution of individual stars to find out about their life-styles. Stars like the Sun live for thousands of millions of years, and only a small minority of all stars show any significant changes in a human lifetime—or even in the entire history of human civilization. Instead, astronomers get a handle on how stars change by studying the great variety of stars in the sky which are accessible to their telescopes and recording instruments. Stars at all stages of evolution from birth to death are seen in our own Galaxy, and once the different stages have been identified it is possible (though by no means simple) to get an overall picture of how a typical star lives out its life. This is rather as if an intelligent extraterrestrial was able to obtain a snapshot of everything happening on Earth at some instant of time. Although this frozen image would be unchanging, it would include information about all stages of human life, from birth to death, and the intelligent alien would be able to work out the human life cycle from studying the evidence of 4,000 million lives, snapped at one instant of time, before him (it?).

One of the key factors that is determined from our astronomical "snapshot" of thousands of stars at one instant of cosmic time is the temperature of each star. This is revealed by spectroscopy, the technique by which light is split into a spectrum (or rainbow) of colors for analysis. Such spectra of the light from stars generally show sharp dark lines (in some cases, bright lines) crossing the spectrum at precise wavelengths. This is a phenomenon familiar from laboratory studies here on Earth, and astrophysicists are able to identify these lines with energy absorbed or emitted by particular atomic elements or with the ionized forms of these atoms. This is the key to taking the temperature of the stars. An atom is ionized if it has lost some of the electrons which balance the positive electric charge of the nucleus, and the number of electrons knocked off from atoms of any particular element depends very precisely on the temperature of the gas in which those atoms are being jostled about. Since each pattern of spectral lines is related to the atom, or ionized atom, which produces it like a fingerprint, studying the pattern of spectral lines in the light from any star gives a good guide to the temperature at the surface of that star.

Stars were originally classified in this way on the basis of the strength of lines of ionized hydrogen (hydrogen lines), in an alphabetical scale starting, logically enough, A, B, C. But when the spectoscopic technique was extended to measuring the strength of lines produced by other atoms as well as those of hydrogen, it gave a more accurate guide to temperatures, and the neat alphabetical

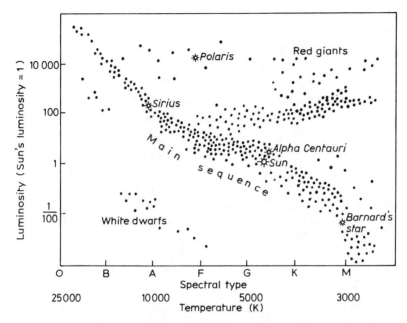

The Hertzprung-Russell diagram, on which the relationships between different types of stars can be seen. Stars to the top left of the main sequence are massive, hot, bright, and short-lived; stars to the bottom right are cool, dim, small, and long-lived. A star like the Sun combines enough warmth to keep a planet like the Earth habitable with a lifetime long enough for organic life to evolve on such a planet. It is no coincidence that creatures like ourselves inhabit a watery planet, with a blue sky, circling a yellow/orange-type G star.

order had to be reshuffled, with a new category, O, hotter than A, added at the beginning. So today the range of stellar classes from hottest to coolest is labeled, in order, O, B, A, F, G, K, M; this might seem difficult to remember, except that some astronomical genius long ago came up with the mnemonic "Oh, Be A Fine Girl, Kiss Me," which nobody who hears it ever forgets!

This range of temperatures goes from about 40,000 K for O stars to 3,000 K for M stars;* cooler stars are also known, but this covers the main range of spectral types. Our own Sun is a G-type

---

* Temperatures given in K are virtually the same as °C at these high temperatures; zero Kelvin is the absolute zero of temperature, at which all thermal motion of atoms and molecules stops. This is −273°C, but the difference between 40,000 degrees and 40,273 degrees is neither here nor there, and O stars might well be described as having surface temperatures

star, with a spectrum dominated by lines of ionized calcium and neutral (nonionized) metals. Like the other categories, the G class is subdivided into ten subclasses numbered 0 to 9, and the Sun is more precisely described as a G2 star, one-fifth of the way between G0 and K0 in terms of temperature and spectrum, although there is no suggestion that stars evolve along this "sequence" of stellar spectral classifications. The surface temperature of the Sun, like other G2 stars, is about 6,000 K.

In terms of color, stars range from blue through white and yellow to red. This too can be explained in terms of temperature, since the hottest stars emit most radiation at the blue end of the spectrum, while cool stars radiate their energy at the red end of the spectrum. Our Sun is a yellow star, roughly in the middle of the band of visible light. This is no coincidence; we have evolved for thousands of millions of years under the light of a G2 star, and our eyes are so adapted that the light from the Sun is in the middle of the band we can see. If we had evolved on a planet circling an O or A star, we would surely have eyes sensitive to blue light, with a "visible spectrum" perhaps ranging into the ultraviolet, and with yellow or red light perhaps as invisible as infrared radiation is to us here on Earth. Creatures living by the light of a K or M star, on the other hand, might have "eyes" very good at "seeing" infrared, but blind to blue light—if, that is, life could ever evolve under the light of either kind of extreme. For, while it is no coincidence that our eyes are adapted to the light of the star at the center of our Solar System, it may also be no coincidence that we live by the light of a G-type star. As we shall shortly see, these relatively quiet but relatively long-lived stars may be ideal homes for the evolution of intelligent life, and the planets circling O and A stars at one extreme, or K and M stars at the other, may be far less hospitable than Planet Earth.

So identifying the temperature of a star gives the first real astronomical "handle" on its nature. The next crucial question is the one of the distances to the stars, something which is not directly revealed by measurements of either temperature or brightness. A

---

around 40,000°C. But the difference is more important for cooler stars, and once we get down to conditions here on Earth, a difference in temperature of 273 degrees is absolutely crucial to life.

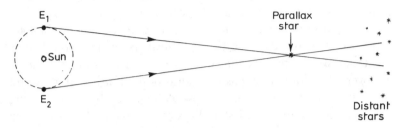

The distance to a nearby star can be determined from the parallax method—by measuring the angle across which the star appears to move when viewed from opposite ends of the Earth's orbit around the Sun.

star might be small and hot, say, and quite nearby, or it might be very big and hot, and a long way away, and still present exactly the same appearance to us in terms of its brightness and spectrum. For a fixed temperature, if the size of a star is doubled its brightness increases fourfold. So getting a second handle, on distance, gives astronomers an idea of the size of stars, as well as their surface temperatures. That, in turn, gives an all-important indication of how hot those stars must be in their *centers* where all the energy that keeps them hot is produced.

STELLAR DISTANCES

The first method of determining stellar distances depends on the way nearby stars appear to shift across the background of more distant stars as the Earth moves in its orbit around the Sun. This is exactly the same effect as the way in which nearby telegraph poles, viewed from a moving car or train, seem to slide past the background of distant mountains, or the way the Moon seems to "keep up" with you when you are out walking at night while nearby houses or trees seem to move relative to the Moon. The orbit of the Earth around the Sun has a diameter of about 300 million kilometers, which gives us a baseline for measuring the parallax effect; the closer a star is, the bigger the shift observed over a year will be. In fact, although astronomers make their measurements over a year, taking full advantage of the 300 million-kilometer baseline, they define the measurements they make in this way in terms of the *radius* of the Earth's orbit, simply dividing the angular shift of a nearby star over a full year by two. In this way, they define the unit of distance called the parsec, mentioned in Chapter One, which is the distance to a mythical star which would show a parallax shift of two seconds of arc over a year, or one second of arc over a baseline as long as the radius of the Earth's orbit around the Sun. No star, in fact, is close enough to show such a big parallax shift. The nearest star, Proxima Centauri, has a parallax of 0.765 arc seconds, and is about 4¼ light-years away (one parsec, remember, is roughly 3.26 light-years). The Northern Pole Star, Polaris, is just over 200 parsecs away, nearly 700 light-years, and this means that this particular star must be 10,000 times as bright as the Sun to shine so brightly in the night sky here on Earth.

The parallax technique, involving incredibly precise measurements of angular shifts much smaller than anything that could be detected by the human eye, can be used to determine distances of stars out to about 300 parsecs, or 1,000 light-years—which includes the vast majority of the stars that we can see with the naked eye. Distances to more remote stars can be worked out in several ways. First, there is another geometrical technique, the moving-cluster method, which depends on the fact that stars moving in a group through space look as if they are converging on a point; this depends on measurements of photographs taken over many years, to provide an accurate indication of the speed with which the cluster is moving across our line of sight. By the time both these geometrical techniques have been applied to the limit, astronomers have accurate distance to thousands of known stars, and from this evidence they can get a very good idea of how bright stars of particular spectral types really are. So any star of the same spectral type can be assumed to have a certain brightness, and its apparent brightness indicates its distance. In addition, some types of variable star, the Cepheid variables and RR Lyrae variables in particular, show very regular fluctuations in brightness, which observations of nearby stars studied geometrically show to be related to the average brightness. These and other techniques push out the astronomers' "measuring stick" across space, to the ends of our own Galaxy and then over into other galaxies. As I have mentioned, these techniques are crucially important for cosmologists studying the behavior of distant galaxies. But we shall have to gloss over them here and return to the theme of how knowledge of both temperature and distance is sufficient to unlock the secrets of stellar evolution.

Distance measurements tell us how bright a star really is; spectroscopy tells us how hot it is at its surface. Putting the two together, it is possible to relate brightness to temperature in a kind of graph known as a Hertzprung-Russell diagram, after the Dane Ejnar Hertzprung and the American Henry Russell, who independently hit on the technique early in the twentieth century. Such a plot shows a broad band, on which most stars lie, running from the top left of the diagram (corresponding to very hot, very bright stars) to the bottom right (corresponding to cool, dim stars). In addition, there is a "branch" from the main sequence, extending up to the right of the diagram, corresponding to bright but cool stars, and a scattering down in the bottom left-hand corner, representing hot but faint stars. All of these features can be simply explained by the laws of physics, and the explanation gives an important guide to how stars change as they age.

STELLAR LIFETIMES

Taking the main sequence first, it seems clear that the hottest, brightest stars are the most massive. The more mass a star has, the stronger is the inward tug of its self-gravity, and the more energy must be created in the heart of the star to hold it up. This means burning the nuclear "fuel" at a profligate rate, pouring out energy and keeping the surface of the star hot and bright. Such stars can only have relatively short lifetimes on the main sequence, before their fuel is exhausted and drastic changes occur. At the other extreme, the cool, dim stars are very small, with weaker gravity, and only a modest amount of energy production in their hearts. They can last for a very long time without changing. And in the middle there are stars like the Sun (and including the Sun), moderate in size and rate of energy production, moderately hot, and not very big. Bright O stars may be forty times as massive as the Sun (or more), 500,000 times as bright as the Sun, and with sizes (in terms of radius) eighteen to twenty times that of the Sun. Dim M stars, on the other hand, may shine with only one-sixteenth or less of the Sun's brightness, having half the Sun's mass and just over half the Sun's radius.

Getting off the main sequence, the cool but bright stars are identified as giants, stars which are very much larger than the Sun, so that even though each square meter of the stellar surface shines less brightly than each square meter of the Sun, the overall effect is to produce more radiation. By contrast, the hot but dim stars must be dwarfs, smaller than the Sun but with each square meter of surface shining brighter than the corresponding area of our Sun. Putting together the evidence gathered by a whole generation of astronomers—and glossing over the detailed story of how they got the evidence and refined their theories into the form we have today—all of these features are explained in terms of the evolution of stars of different mass.

When the Sun formed from a large cloud of collapsing gas, it was undoubtedly large and cool, so it started in the region of the Hertzprung-Russell diagram above and to the right of the main sequence. As the forming Sun contracted, it got hotter and brighter, moving to the left on the diagram, and the point where this "evolutionary track" met the main sequence corresponds to the balance struck between the nuclear fires inside the Sun, providing heat energy to hold it up, and gravity, still trying to pull it into a tighter ball. The Sun will stay more or less at this same point on the main sequence for a total of about 10,000 million years, roughly half of which have passed already. The mass of a star decides how hot it must be inside to withstand its own gravitational pull, and this fixes the temperature of the star and its brightness, so that each mass corresponds to one possible point on the main sequence. A more massive star, bound more tightly by gravity and correspondingly hotter, lives out its short, main-sequence life in the top left of the Hertzprung-Russell diagram. And although a star with twenty times the Sun's mass will have twenty times as much nuclear fuel to burn, it will be 10,000 times as bright as the Sun, using up its fuel in less than 1,000 million years. Polaris is such a star. But, hardly surprisingly, less than 1 percent of the stars in our neighborhood in space are hotter than class F in the alphabetic sequence, while the quiet, long-lived stars of category M and below make up almost 75 percent of the population of the disk of the Milky Way. This raises interesting implications for the occurrence of life in our Galaxy and in the whole Universe, since it is only during its life on the main sequence that a star behaves in a steady and reliable fashion for long enough to allow intelligent life to evolve on any planets it may possess. Here on Earth, where conditions seem to be very good for life (as we shall see later), it has, even so, taken more than 4,000 million years of quiet, main-sequence existence for intelligent life to emerge.* The flashy, short-lived O, B, and A stars look, from this perspective, very unlikely candidates as center-pieces for solar systems containing life. Even most of the F category would start to show dramatic changes over about the same sort of time which it has taken for us to develop here on Earth. On the other hand, while the dim M stars are very long-lived, giving plenty of time for life to evolve on any planets they possess, they are very weak energy sources—and life certainly needs energy. A cold ember at the heart of a planetary system might never provide sufficient input of energy for the kind of life we know to get a grip on a planet.

In trying to make generalizations about the nature of the Universe, and the prospects of finding life in different parts of the Universe, human beings always have to be wary of assuming that life must indeed be "as we know it." This very straightforward evidence from the Hetzprung-Russell diagram and the insight it gives us to stellar evolution does, however, strongly hint that it is no coincidence that intelligent life has evolved on a planet circling a G2 star. As this story of the origins of mankind within the evolving Universe now begins to focus closer to home, this is an important signpost to the road ahead. Increasingly, now, it seems that life "as we know it" is neither a freak event in the Universe nor simply one variety among many, but rather the inevitable product of the kind of Universe we live in. And this approach suggests that the best stars for life to evolve under are the ones which strike a balance be-

---

* The Sun may not have been *perfectly* steady throughout the past 4,500 million years, and its minor fluctuations may have been very important to life on Earth. However, by the standards of its own changes both before and after its main-sequence "life," the Sun has indeed been relatively steady for all this time.

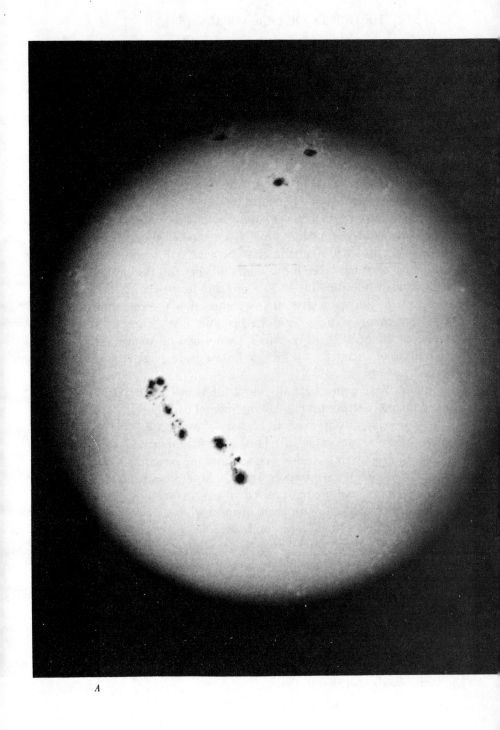

A

Our Sun is an ordinary, unspectacular star, but it is far from being completely quiet. Dark spots on the surface come and go over a cycle roughly 11 years long, and photograph A (taken by Paul Roques of the Griffith Observatory) shows a group of sunspots visible to the naked eye on April 8, 1980. When the Sun is more "spotty," it is also more active in other ways, producing great flaring outbursts like the one shown in this NASA/Skylab picture (B). To get an idea of scale, remember that the diameter of the whole Sun is 109 times the diameter of the Earth. (*NASA Photo*)

B

tween pouring out energy for life to feed off and lasting for a long time to give life a chance to evolve. That balance occurs in the middle of the main sequence, and simply by recognizing the need for a trade-off between longevity and energy the search for life must focus on the planets of G-type stars. As far as we know, there is nothing special about the Sun as a G-type star, nor about its family of planets. This suggests that there may very well be a lot of planets like our own, many of them rich with life more or less "as we know it," although surely rather more different from us than we are from a crab or a kangaroo.

After the Sun has finished its life on the main sequence, it will go through many changes. These will not happen for 5,000 million years, and might seem of little relevance to the story of human origins. But other stars—especially more massive stars—have trodden the same path before, and it is thanks to them that we exist today. The origin and life of the Solar System depend on the deaths of stars that have gone before, and to understand this we need to understand, in outline at least, the processes which go on inside stars to generate their energy output.

STAR DEATH

Stellar energy comes from nuclear reactions, the fusion of atomic nuclei in the hot, central regions of a star. Like the primeval fireball of the Big Bang but on a much less dramatic scale, the material inside a star is ionized completely, with positively charged nuclei and negative electrons making up a sea of particles called a plasma. Originally, the heat of a stellar interior comes from the gravitational energy released by the collapsing gas cloud out of which the star formed; but once the temperature is high enough, the light nuclei can be literally fused together to make the nuclei of heavier elements. In the process, a little mass is converted into energy, which keeps the pot boiling and stabilizes the star against further contraction—as long as the fuel lasts. Since most of the matter in the original gas cloud was primeval hydrogen, the process starts with the fusion of hydrogen nuclei (single protons) to make helium nuclei. In stars like the Sun, this happens through straightforward proton-proton interactions, building up stable nuclei of helium-4 (each containing two protons and two neutrons) from interactions involving four protons and the release of two positrons (the positively charged counterparts to electrons). The net effect, turning four hydrogen nuclei into one helium nucleus, takes place all the time, with the conversion of 600 million tons of hydrogen to helium every second being required to produce the energy output of the Sun today.*

We know how hot the middle of the Sun must be from the amount of energy which escapes from the surface and from the laws of physics, which tell us how much energy is needed to "hold up" a star of one solar mass. For the Sun, the internal temperature is close to 15 million K, and the nuclear burning process is dominated by the proton-proton chain. Bigger stars, with central temperatures above 15 million K, are kept hot mainly by another process, the carbon-nitrogen-oxygen cycle, which involves interactions of all those nuclei as well as hydrogen and helium, but has the same net effect of converting hydrogen nuclei into helium nuclei and releasing energy along the way. The problems—and the end of quiet, main-sequence respectability—come for stars when there is no hydrogen left to convert into helium in the center of the star.

When this happens, there is still hydrogen left in the outer layers of the star, so at first the hydrogen "burning" starts to spread outward in a shell, surrounding a central core almost entirely composed of helium nuclei. The core contracts and gets hotter,

---

* It is, in fact, just possible that the Sun is not being maintained by nuclear burning at present. Several lines of evidence suggest that something odd is going on inside the Sun, and has been for perhaps a few million years and certainly for a few thousand years. One key clue comes from attempts to detect particles called neutrinos, which should be produced in great quantities if the nuclear fusion processes are operating in the Sun as theory predicts. No more than one-third of the expected flood of neutrinos has been detected, and the simplest explanation is that the nuclear burning has temporarily "switched off." This is borne out by the dramatic discovery, made by Dr. Jack Eddy from studies of observations going back a hundred years or more, that the Sun may be shrinking at a slow rate today. This is exactly what we would expect of a star if nuclear burning were switched off—the star starts to collapse under gravity, but maintains its brightness as gravitational energy is converted into heat. Such a hiccup could last for 100,000 years without seriously affecting conditions here on Earth, and set against the timescale of the Sun's main-sequence life, 10,000 million years, this really is no more than a hiccup.

The implications of even a small disturbance of this kind for human life, however, could be profound. They are discussed in detail in my book *The Death of the Sun* (New York: Delacorte, 1980), which also covers in more detail the nuclear fusion reactions that occur inside stars when they are not hiccuping.

since gravitational energy is released, while the outer layers of the star become swollen by the greater heat escaping from the center, swelling up so that the star becomes a red giant, above and to the right of the main sequence in the Hertzprung-Russell diagram. When our Sun reaches this stage, it will be a thousand times as bright as it is today, and with a radius 100 times bigger, completely engulfing the Earth. But there will be plenty of warning for anyone still around in 5,000 million years to notice. The expansion takes hundreds of millions of years; but eventually, after this phase of its life, even more exciting changes begin to take place inside the star.

Just what happens next depends on the mass of the star. For anything less than about two solar masses, the compact and increasingly hot helium core suddenly reaches a stage where it is hot enough for helium nuclei to fuse into those of heavier elements, and helium "burning" begins explosively in a stage of evolution known as the helium flash.* After the flash, the star settles down again, with a helium-burning core still surrounded by a hydrogen-burning shell. More massive stars, however, achieve the change without the drama of a helium flash, adjusting themselves more gradually to the combination of hydrogen and helium burning which dominates the energy production of virtually every red giant visible in the sky. At the helium-burning heart of such a star, temperatures reach 100 million K as helium is converted into carbon and oxygen.

For some stars, this is the end of the story. When the hydrogen and helium are both used up, the star simply cools off and contracts down into a small, cold ball, a dim dwarf star in the bottom left of the Hertzprung-Russell diagram. Such a white dwarf (eventually cooling to a black dwarf) may be as massive as the Sun but only the size of the Earth, held up not by heat from the interior anymore but by the simple physical strength of the atomic nuclei packed together in what has become, in effect, one huge crystal of

* All of this we know not from observation alone but a combination of observation and theory. Different kinds of stars can only exist as observed if their interiors have certain temperatures and pressures; physical laws determined on Earth tell us what happens to matter inside stars under those temperatures and pressures, and the theorists then compute mathematical "models" of what will happen next, and how stars got to be that way in the first place.

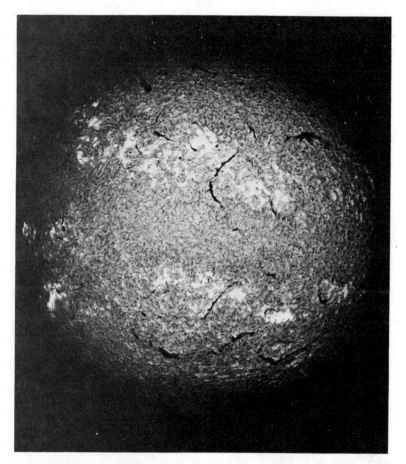

One way of showing the Sun's activity is to make a photograph using only the light emitted by hydrogen ("hydrogen alpha" light). This picture, obtained by Pennsylvania State University observers on May 29, 1968, shows white swirls in the regions where flares are likely to occur and narrow black lines related to features known as prominences. *Credit: Pennsylvania State University*

carbon with a frozen skin of hydrogen and oxygen—a cinder covered in ice.

But for massive stars, the helium-burning phase may be followed in turn by further core contraction, hotting up the interior to switch on carbon, oxygen, or even silicon "burning." An old, massive giant star may contain as many as five such nuclear burning shells, with iron being built up at the center. But when this stage of evolution is complete, and the iron core collapses and hots up, it has come to the end of the road. To make still heavier nuclei from iron-56 you have to put energy *in*—you don't get anything out. Indeed, very massive nuclei, such as those of uranium and plutonium, tend to split up, releasing energy in the process as they fragment into lighter nuclei. This is the fission process of an atomic bomb. So where do such heavy elements come from in the first place?

Here, the modeling of the theorists becomes less certain. Astrophysicists think they know roughly what happens, but they cannot, as yet, spell out the details. It is clear that as the iron core collapses and releases its gravitational energy there must be a vast explosion which rips the giant star apart, and it is thought that the outpouring of all this gravitational energy is sufficient to build up heavy elements, including uranium, along the way. Even with some energy going into nuclear reactions in this way, though, enough is left to produce the most spectacular stellar events in the Universe—the supernovae. For a brief period, a star which dies in a supernova explosion may shine as brightly as all the rest of the stars in a galaxy put together—and such explosions occur once every few decades among the stars of a galaxy like our Milky Way. Now we return to the link between star deaths and star births, recalling that our Solar System seems to have been created by the squeezing effect of one or more supernovae on a cloud of gas in space. But there is still more to this phoenixlike story of rebirth.

The exploding debris of a supernova scatters material far across space in an expanding cloud of gas. But this material is very different from the collapsing cloud out of which the star had originally formed. Originally almost entirely hydrogen and helium, the star, over its lifetime, has built up a mixture of every other known element, including the heavy elements literally stuck together in its death throes. All of these elements—including carbon, oxygen, iron, uranium, and everything else we know on Earth—spread out to enrich the interstellar medium, and eventually to form part of new clouds of gas which are in turn squeezed and compressed to the point where they collapse to form new stars and planets. Quite literally, everything on Earth except hydrogen and helium (and quite possibly most of that too) has been processed in the inside of at least one star. The atoms in our bodies, in this book, in the ground we walk on, and in the air that we breathe have all been through the stellar cauldron. And without preceding generations of massive stars and supernova explosions, I wouldn't be here writing this book and you wouldn't be there reading it.

But, of course, not all of the material in the exploding supernova gets recycled. The heart of the iron core gets squeezed still further while the outer layers of the star are being blasted outward, and remains as a remnant at the heart of the expanding cloud of material. This may be squeezed so much that even the strength of atomic nuclei, which holds dwarf stars up against the pull of gravity, cannot resist, and like running a movie film backward, the complexity of heavy nuclei built up so laboriously by nuclear fusion is destroyed as the nuclei dissolve into a sea of neutrons. In such a state, the whole stellar remnant collapses into the equivalent of one giant atomic nucleus, a neutron star. Where a white dwarf as massive as our Sun might be the size of the Earth, a neutron star with one solar mass would be as small as the island of Manhattan.

Even this is not quite the end of the story. For if the remnant has more than a couple of solar masses of material, even the strength of neutrons cannot hold it up anymore against the pull of gravity. After a drawn-out struggle which has lasted since the original gas cloud began to collapse, opposed by nuclear fusion and with the last fling of a supernova explosion, gravity at last wins its battle with matter, and the remnant is squeezed out of existence altogether, leaving behind a black hole. Neutron stars and black holes are among the most exciting discoveries of modern astronomy, and provide the powerhouse for energetic celestial objects such as pulsars, X-ray stars, and (on a rather grander scale) quasars and radio galaxies. But these exotica have no direct bearing on the story of the origin of life on Earth. Armed now with the knowledge of how the lives and deaths of massive stars relates to the origin of our own Sun and Solar System, we can at last bring our story down to Earth, and look at how our home planet originated and developed into just that—a home for life.

4 THE ORIGIN OF THE EARTH

The most important thing we know about the origin of the Earth is that our home planet formed at the same time as the Sun, with the rest of the Solar System, from a collapsing cloud of gas in interstellar space. But the story we are following in this book is primarily the story of the origin of life on Earth in general, and of human life in particular. Our form of life needs a planet with a star like the Sun, which inevitably was the product of a galaxy which itself formed from fireball fluctuations in the expanding Universe. At each stage of the story, it has been necessary to focus down to a finer scale, and the time has come to do so again. Having looked in breadth at the formation of the whole Solar System, I want now to look at the origin of the Earth *as we know it*, as a home for human life. This means looking not so much at the whole Earth, including its deep interior, but particularly at the skin of the Earth, the crust, oceans, and atmosphere which provide the home for life on Earth.

This concentration on the Earth's skin—sometimes called the biosphere—is not only appropriate but necessary, since in fact very little is known about the details of the Earth's interior. In many ways, it is more difficult to find out what goes on deep beneath our feet than it is to find out about the far reaches of the Universe, since at least we can see light from the far reaches of the Universe! The following outline of what we do know about the Earth's interior and how it formed is therefore superficial, and the ideas described are likely to be revised as our understanding improves over the next decade or so. But, after all, before I turn my attention to the crust and biosphere, at least passing mention should be made of what lies beneath the crust.

INSIDE THE EARTH

The cloud of dust and gas within the Solar System which collapsed to form Planet Earth was probably composed chiefly of silicon compounds, iron oxides, and magnesium oxides, with just a trace of all the other chemical elements we now find in our terrestrial environment. This debris itself represented just a tiny fraction of heavier material left behind in the inner Solar System, with lighter elements driven outward by the activity of the Sun as it formed. As the Earth formed, it warmed up through three processes. First, as particles collided with one another and stuck together, their energy of motion (kinetic energy) was converted into heat. This continued well after the planet had grown to a respectable size, with the "particles" doing the colliding then being large meteorites smashing into the surface of the planet—as still happens occasionally today. Then, as a planet proper began to grow by this accretion, the interior was pressed ever tighter by the increasing weight of material above. Just as a gas cloud or star warms up as it contracts, releasing gravitational potential energy as heat, so did the proto-Earth. Unlike a gas cloud or star, however, the heat from the interior could not escape easily into space, but was largely trapped by the solid barrier of the crust above, helping to keep the interior of the Earth hot to the present day. And, thirdly, some of the elements in the mixture of heavy materials that made up the core of the young planet were radioactive—the heavy products of the nearby supernova that triggered the birth of the Solar System—and have since been decaying, giving up energy of nuclear fission as a further contribution to the heat of the Earth's interior. The radioactive heat was produced in greater quantities when the Earth was young, since many, many atoms have now decayed and relatively few are left to contribute to the warmth of the Earth today. Almost certainly this early radioactive heating was a major factor in melting the iron-rich Earth, perhaps a few hundred million years after a planet began to form from the cloud of material circling the proto-Sun. But because the melting temperature of a solid is increased if the solid is under pressure, melting did not begin at the center of the Earth, but between 400 and 800 km depth, where the combination of slightly lower pressure than in the middle but distinctly higher temperature than at the surface provided an appropriate balance.

Molten globules of iron, formed in this way, would then have sunk through the material around them toward the center of the Earth, giving up more gravitational energy as heat in the process, with almost the whole of the interior eventually melting. At this stage, heavier elements could settle freely into the deep interior, while lighter elements could float toward the surface. So the Earth began to approach its present structure, with a molten interior rich in iron and a surface layer of lighter material, dominated by silicon. Just as the composition of the whole Earth is not typical of the composition of the cloud from which the Solar System formed, so the material of the crust on which we live is not typical of the material which makes up the bulk of the Earth. But at every stage along the way that separation of elements from one another has occurred, it has followed straightforward physical laws. There is nothing peculiar about the separation of materials which led up to the environment present today at the surface of the Earth, and any planet forming in the same way would end up in a very similar state. Our terrestrial environment is very much a product of the Universe we live in, right back to the fireball of the Big Bang itself.

So today the crust of the Earth contains just 6 percent iron, whereas the planet as a whole contains 35 percent iron, while the crust is made up of 28 percent silicon, which only makes up 15 percent of the mass of the whole planet. Some heavy elements, such as uranium and thorium, never got a chance to settle deep into the Earth's interior, because they easily combine with light elements such as oxygen and silicon, making light compounds (oxides and silicates) which happily floated up into the surface layers. One result of this is that after the initial melting a great deal of the radioactive heat was produced in the surface layers, where it could easily be radiated away into space. Indeed, once the surface layers of light material had formed, and very little hot material from the interior was rising to the surface, cooling was able to produce the solid crust of the Earth that established more or less the kind of surface we know today, but without the atmosphere or oceans we now know, and certainly without life. Instead, as the crust solidified, the Earth's surface was covered with hot rock and a wealth of active volcanoes, pouring out both molten rock (lava) and gases. These gases played a crucial part in the subsequent evolution of the surface of our planet, as we shall shortly see. But since it was at the stage of cooling and solidifying that the main feature of the crust, its division into continents and what were to become ocean basins, must have occurred, first let's take a last look at how the structure of our planet, from crust to deep interior, stabilized.

The continents are made of light rock, chiefly granite, while the material of the crust beneath the oceans, chiefly basalt, is rather heavier, so that to the very end the lighter materials were rising higher than the denser materials, in line with the simple laws of physics. We can even put a date on when all this happened, since the oldest rocks known on Earth are 3,900 million years old, and the age of the Solar System is about 4,500 million years. So, in round terms, it took about 600 million years for the proto-Earth to develop a hot, molten interior, for "differentiation" of light and heavy material to take place, and for the crust to begin to solidify—just about 13 percent of the history of our planet to date.

What was produced after all that activity was a layered planet, with several shells wrapped around one another. And what we know about this layering today is a result of the study of earthquake waves, which pass right through the Earth and can be monitored by seismic detectors. Seismologists measure the time it takes the waves to pass through the Earth from a known earthquake to different points around the globe, and this tells them something about the nature of the material inside the Earth that the waves

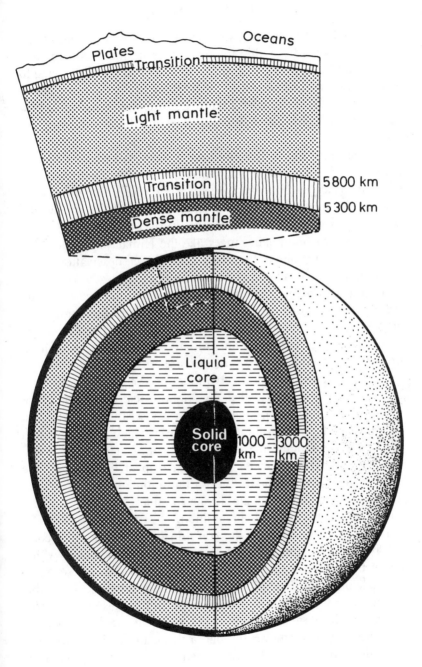

Oceans

Plates

Transition

Light mantle

Transition

Dense mantle

5 800 km

5 300 km

Liquid core

Solid core

1000 km

3000 km

The layered structure of the Earth today.

have passed through. In all honesty, unraveling a picture of the Earth's structure from this information is as much an art as a science, and one eminent geophysicist described it as like trying to analyze the structure of a piano by listening to the noise it made as it was pushed down a flight of stairs. It is therefore remarkable how much seismologists have determined about the structure of the Earth.

Starting from the surface, we have a rocky ball with an average radius of 6,371 km, slightly flattened at the poles. The outer shell or crust makes up only 0.6 percent of the volume of the planet, and its thickness varies from a fairly uniform 5 km below the oceans to 35 km under flat continental surfaces and as much as 80 km under great mountain ranges like the Himalayas. At least we know something about the crust from direct observation. It is made up mainly of silicon and aluminum, plus oxygen, combined in various compounds to make granite rocks, solidified lava, and sedimentary deposits. Overall, the name given to the continental crust material is sial, from the first letters of silicon and aluminum; because oceanic crust also contains large amounts of magnesium, in various compounds, it is sometimes referred to as sima.

Below the crust there is a very pronounced change in structure, marked by a boundary called the Mohorovičić discontinuity (or simply Moho for short), after the Yugoslav seismologist who first identified it. Below lies the mantle, a layer which goes down to a depth of 2,900 km and makes up more than 82 percent of the volume of the Earth. It seems to be made up of three zones, the upper mantle about 370 km thick, a transition zone some 600 km thick, and the lower mantle 1,900 km thick—but very little is known about its chemical composition.

Deeper still we come to the core, divided into the outer core, 2,100 km thick, and the inner core, 1,370 km in radius, which lies at the center of the Earth. This is the region which seems to be made of liquid iron, with a mixing of something else—probably sulfur. One of the most important results of the existence of a core of molten iron in the Earth is the magnetic field it produces. When a conductor moves through a magnetic field, electric currents flow, and this must have happened early in the formation of the Solar System when the newly molten core of the Earth interacted with stray magnetic fields from the forming Sun. Once an electric current is set up in this way, it produces its own magnetic fields, and, fed by the rotation energy of the spinning Earth, the result is a self-acting dynamo which has persisted to the present day. This feature of our planet may have played an important part in the history of life, and it has also provided us with very useful magnetic evidence, frozen into the rocks of the crust, of how the surface layers have changed since they solidified 3,900 million years ago. Before we come to this, however, there is still an important subtlety to describe about the surface layers themselves, and, of course, the story of the atmosphere and its evolution.

As well as the basic division of the Earth into core, mantle, and crust, a division which depends mainly on chemical composition, there are subdivisions of the surface layers which seem to depend simply on physical difference (pressure and temperature) and not on chemistry. From 75 km depth to about 250 km, there is a layer which has a slightly lower seismic-wave velocity than the layers just above and just below it. This is not quite a liquid layer, but a region of weakness where the mantle is partly molten, like a pail of slushy, melting ice. This is called the asthenosphere and is described as being like a viscous sea (perhaps thick porridge is a better metaphor than slushy ice!) on which the layer of crust above, the lithosphere, floats. The solid layer below the asthenosphere—the rest of the upper mantle—is called the mesosphere. And the presence of the asthenosphere is all-important in terms of what has happened to the surface, the lithosphere, in the past few thousand million years, since this treacly layer has allowed the solid rocks above to move about, shifting continents around the globe, opening up ocean basins here and squeezing oceans shut there. The asthenoshpere is also the source of new crustal material, in the form of magma which rises to the surface from the interior.

With the story of drifting continents and changing ocean basins, we are almost ready to look at the origin of *man's* Earth, the thin skin of the biosphere in which life can exist. Life as we know it, however, could not get a foothold until the oceans formed and an atmosphere was created. And life on land could not get a foothold until the atmosphere was modified—with the aid of life in the sea—into the oxygen-rich brew that keeps animals, including human animals, living and breathing. The last stage in the birth of the Earth as a home for life was the evolution of the air that we breathe. To explain this, the origin of the atmoshpere, it helps to take one step back into space, to compare the events going on here on Earth early in the history of the Solar System with the events going on on our near neighbors, Venus and Mars.

THE AIR THAT WE BREATHE

At last we have come down from the cosmic viewpoint to the stage where we are dealing with the direct environmental surroundings of life as we know it. Is life—even intelligent life—something that arises almost inevitably on planets that are the "right" distance from the parent star and contain the "right" chemical imgredients? Or is life some kind of cosmic fluke, which has arisen only occasionally—perhaps only once? Until we find life on another planet, we can never be absolutely certain of the answers to these questions. But we do know that life has got a very firm grip on one particular planet, this one. And here on Earth the success of life seems to be intimately connected with an abundance of liquid water—as our term for a region devoid of life (desert) indicates, being the same as our term for a region devoid of water. We have already seen that planets like the Earth, Venus, and Mars are inevitable consequences of the process by which stars form, provided that the prestellar cloud breaks up to give a star plus a family of planets rather than a multiple star system.* If we now make the reasonable guess that life requires a wet planet, then every other factor—temperature, chemical composition, and so on—is automatically taken into account in providing the physical conditions necessary for the presence of large amounts of liquid water. Leaving aside entirely the question of life as we *don't* know it (and

---

* This is an important *caveat*. Only about 15 percent of the stars in the disk of our Milky Way Galaxy are not accompanied by one or more other stars, with the members of the multiple system orbiting about their common center of mass. It is hard to see how planets in such a system could be very pleasant for life, even assuming that they could stay in a stable orbit under the constantly varying gravity pull from two or more "suns," since there would probably be extreme variations in temperature with several furnaces in the system providing heat at different times. So the cautious view is that only 15 percent of the stars in the Milky Way could have planets like our own—but 15 percent of several thousand million still leaves plenty of possible sites for life.

some astronomers have speculated seriously on the possibility of, for example, floating life forms living in the atmosphere of Jupiter), the answer to whether *we* are a fluke product of rare astronomical conditions depends on whether our wet planet is a fluke product of rare astronomical conditions, something simply assessed by looking at the reasons why our near neighbors, Venus and Mars, have failed to become wet planets.

Like the Earth, these are small, rocky planets, as any planets in the inner part of the Solar System must be. The nature of the atmosphere and oceans that develop over small, rocky planets of this kind depends on one thing only—the exact equilibrium temperature of the rocky surface, which is related to the exact distance from the central furnace of the Sun.

Venus, so nearly the Earth's twin planet in many ways, has a thick blanket of atmosphere, rich in carbon dioxide and laced with sulfuric acid clouds, which traps solar heat through the so-called greenhouse effect and raises the temperature at the surface to 500°C, way above the boiling point of water; Mars, on the other hand, has a thin carbon dioxide atmosphere, only a small greenhouse effect to help warm it, and a surface temperature too cold for water to exist as a liquid or for rain to fall. The Earth alone of the three has an atmospheric blanket which is just right to keep the surface of our planet in the temperature range between the melting point of ice and the boiling point of water. In each case, the present "environment" has grown up from the gaseous material released from the interiors of the three planets—any primordial atmosphere (perhaps dominated by methane and ammonia) must have been blown away by the Sun's erratic fluctuations as it settled down to a stable nuclear burning state, leaving bare rocky planets to grow their own atmospheres by outgassing, including volcanic activity and the vaporization of surface material when large meteorites hit the surface.

This understanding of the origins of the atmosphere of the Earth is still fairly new. It used to be thought that as the Earth settled down, some 4 thousand million years ago, it would have had an original atmosphere of methane, ammonia, and similar compounds, rather like the atmospheres of the giant planets today. This idea was tied in with the search for the origins of life, since

The battered face of Mercury, shown here in pictures from NASA's Mariner 10 spacecraft, shows more clearly than any other object in the Solar System how planets are made by the accretion of smaller lumps of material that collide, stick together, and build up into planet-sized lumps.

laboratory experiments in the middle of the twentieth century had shown that when gases like methane and ammonia were mixed with water in a sealed tube, and electric sparks or ultraviolet radiation were passed through the tube, then molecules regarded as the precursors of life could form. The early Earth was certainly bathed in ultraviolet radiation from the Sun, and the early atmosphere probably had plenty of sparks in the form of lightning, so the guess was made that it also contained methane and ammonia to set life on its way.

But more recently other experiments have shown that the prebiotic molecules can also be built up in test-tube "atmospheres" rich in carbon dioxide, removing the need to invoke a methane/ammonia atmosphere around the early Earth. Some astronomers, as we shall see in the next chapter, even argue that advanced precursors of life, or even living molecules, are present in interstellar clouds, and could "seed" a new planet, although that extreme view is not wholeheartedly supported by everyone yet. Either way, it is clear that a chiefly carbon dioxide atmosphere plus water is fine for getting life started on Earth. This is a very satisfactory discovery, since volcanic outgassing on Earth today produces plenty of carbon dioxide and water plus sufficient nitrogen, in various compounds, to have built up this major constituent of the atmosphere since the solid Earth formed. There is no reason to believe that the outgassing processes have changed much since the Earth formed, so that we should expect all the rocky planets to produce atmospheres rich in carbon dioxide and water. Mercury, of course, is so close to the Sun that any atmosphere it tries to grow is soon stripped off by the heat; but Venus and Mars both have carbon dioxide atmospheres, which very nicely agrees with the present ideas on how atmospheres formed. They, however, seem to have lost their water, while Earth has lost the carbon dioxide (or, rather, turned it into other things). These differences are neatly explained by the differences in the distances of the three planets from the Sun.

Just about the only thing that simple physics can tell us about the surface of a rocky planet, bare of atmosphere, at a certain distance from the Sun, is its temperature. And that, it turns out, is all we need to know to decide if a rocky planet will be wet or dry! The

Viewed from space, as in this METEOSAT picture, the Earth is strikingly different from the other inner planets of the Solar System, with clouds of water
vapor in the atmosphere and oceans of water surrounding the continents.

equilibrium temperature of a rocky body orbiting the Sun depends on the amount of heat coming in from the Sun (which depends on distance) and the amount being radiated away into space (which depends on the temperature of the rocks above absolute zero, in K). As long as the rocks of the three planets are much the same composition, which they are, the equilibrium temperature depends on the square root of distance from the Sun, so that Mars, 1.52 times as far from the Sun as the Earth (on the average), would have an equilibrium temperature 1.234 times lower than the Earth if neither planet had any atmosphere.*

For Venus, the stable temperature at which incoming solar heat was balanced by outgoing infrared heat from the rocks was 87°C (360 K) before outgassing got underway. Both carbon dioxide and water vapor released by the outgassing would stay as vapor under those conditions, building up a blanket around the planet. And both carbon dioxide and water vapor are very good at trapping infrared radiation, so that the temperature at the surface of Venus was forced upward until a new equilibrium was reached, way above the boiling point of water, building toward the conditions we find on the planet today.

On Mars, things were very different. With an original equilibrium temperature 30° below zero on the Celsius scale (243 K), water could not even melt, let alone evaporate, and although a thin carbon dioxide atmosphere was produced and does contribute a modest greenhouse effect, this is not enough to make Mars a wet planet today. But it is touch and go—Mars is on the very edge of being a wet planet, and could probably have developed much more like the Earth if it had only been slightly more massive, when its extra gravitational attraction would have helped it to keep a thick enough carbon dioxide atmosphere to provide the essential greenhouse effect warming. The *Mariner* and *Viking* pictures of Mars, which show many features which seem to have been carved by some flowing liquid, including dried-up riverbeds, suggest to some astronomers that Mars did indeed have a thicker atmosphere and running water long ago, and that its cold, dry state today developed as the warming blanket of air slowly leaked away into

* For obvious reasons, this factor of 1.234 is one of the useless facts that sticks in the memory!

space. We know that the rivers, if rivers they are, cannot be new features because the surface of Mars is pitted and marked by many meteorite craters. Some of these craters overlap the riverbeds, which shows that they have struck since the water stopped flowing. And since we have a good idea how many meteorites strike Mars (or rather, how *few* strike it), it is simple to calculate that the riverbeds have been dry for a thousand million years or more, and perhaps for half the age of the Solar System.

The flowering of Mars must have been a very short-lived event early in the planet's youth. It is just possible, however, that the planet could be made wet again, since if the polar caps could be unfrozen they would contribute water and carbon dioxide to thicken the atmosphere again. A giant meteorite might strike the surface and do the job; or subtle changes in the orbit of Mars around the Sun might change its seasons and warm the poles; or, just maybe, in a few hundred years' time our descendants from Earth may undertake to make Mars habitable for man. The important point in terms of the prospects of finding life elsewhere, however, is not that Mars failed to make the grade as a wet planet, but that it very nearly succeeded.

On Earth the original surface temperature was about 27°C (300 K, which is 1.234×243), or, in round terms, 25°C, since, to be honest, none of these estimates are better than a couple of degrees either way. This was high enough for liquid water to flow, but not so high that enormous quantities of water vapor got into the atmosphere and helped the carbon dioxide to produce a runaway greenhouse effect. Quite the reverse—the warm waters of the growing oceans actually dissolved carbon dioxide out of the atmosphere, reducing the greenhouse effect, while the shiny white clouds of water floating in the new atmosphere reflected away heat from the Sun, tilting the temperature balance in favor of a slight cooling. The result was that the surface temperature of the Earth *with* an atmosphere settled down to an average around 15°C, where it has remained ever since. The two balancing factors of reflectivity (albedo) and greenhouse effect even provide a kind of natural thermostat which maintains the surface temperature of the Earth in a fairly narrow range, as a simple imaginary "experiment" indicates.

Although Venus is obscured by clouds (which contain sulfuric acid), NASA's Pioneer spacecraft, in orbit around Venus, has mapped the surface by radar, and artists have been able to produce pictures of how the surface of the planet would look without the obscuring clouds. The radar mapping shows that, like the Earth, Venus is divided into continental regions (in this case, with three main continents) and lower-lying plains covering most of the planet's surface, rather like the sea floor of the Earth without any sea in it. The measurements are even accurate enough to identify great rift valleys on Venus, and all of this suggests a planet where the "geophysics" is rather like that of the Earth, complete with plate tectonics, continental drift and "seafloor" spreading. The differences between Earth and Venus are largely skin deep—the difference in atmosphere resulting from Venus being slightly closer to the Sun and enabling the greenhouse effect to build up to runaway proportions. *Credit: NASA*

The surface of Mars today seems to be a dry desert, as pictures from the Viking landers show. But the photographs taken from orbit suggest that water once flowed across the surface of Mars. The overall evidence is that the planet just missed having an atmosphere thick enough to sustain a greenhouse-effect warming and keep its water in liquid form and that the stream channels are very old features indeed. Had Mars been a little bigger or a little closer to the Sun, or had the Sun been a little warmer, there might have been two planets in our Solar System, Earth and Mars, suitable for life as we know it. *Credit: NASA*

IMPLICATIONS FOR LIFE

Suppose the Earth warmed up a little, as it might begin to do if the Sun itself warmed up by a small amount. The first thing that would happen would be evaporation of more water from the oceans, making more clouds in the sky, which would block more of the Sun's heat and help to minimize the warming effect! Or suppose the Sun cooled. With less evaporation, there would be fewer clouds in the sky, and a bigger proportion of the incoming heat would get through to the ground. So the stage was set for life—a wet planet with a stable atmosphere, keeping the temperature range within narrow limits so that the planet neither froze completely nor fried. With present understanding (including those experiments passing electric sparks through bottles of water and carbon dioxide) it is difficult to imagine how life could have failed to get a grip on such a planet, and from here on the story of the evolution of the Earth's atmosphere is inextricably linked with the story of life on Earth. Very briefly, although ultraviolet radiation from the Sun would have kept the surface of the land sterile even after the oceans formed, life could develop under the cover of the seas. The early life forms found oxygen poisonous, a dangerous waste product of living which they had to get rid of. But by a couple of thousand million years ago, oxygen was building up in the atmosphere, where chemical reactions, stimulated by the Sun's radiation (photochemical reactions), built up a layer of ozone (the triatomic form of oxygen) which blocked out the sterilizing solar ultraviolet radiation. Under this protecting filter, life could move out of the seas and onto the land, while new forms of life had developed which thrived on oxygen. The original oxygen producers polluted themselves out of existence, while their descendants adapted into carbon dioxide–breathing plants and oxygen-breathing animals. A new kind of balance was struck, with life itself part of the "feedback," and the Earth became recognizable as a living planet.

A surprising number of people seem to take the story as I have outlined it here as indicating that life may be very rare. They argue that there is no evidence that life exists on either Mars or Venus, and that Earth was lucky to escape the twin traps of the runaway greenhouse and the frozen desert. Our planet is roughly in the middle of the band of orbits around the Sun where rocky planets have the right surface temperature to form a stable atmosphere with running water and oceans; Venus is too close to the Sun and Mars just too far away. With a slightly different orbit, they say, we would either freeze or fry.

But if Earth *were* in a different orbit, would that leave our present orbit empty? If our planet were out in the orbit of Mars, then Venus might have been in our present orbit, and would have developed along the lines Earth actually followed! A planet as massive as Earth in the orbit of Mars would retain enough atmosphere to keep warm and wet by the greenhouse effect, and our Sun would be orbited by two living planets, not one. Of course, planets can't be moved from orbit to orbit, and there are good reasons why Mars is lighter and smaller than Earth, as we have seen. But look at the situation from another point of view.

Suppose the Sun were just a little bit hotter. Then our planet Earth would still be in the life zone, but the zone would now also embrace Mars; with a slightly cooler central star, both Venus and Earth could be wet, even if Mars froze completely. To my way of thinking, it seems that we are *un*lucky in this Solar System to have just one wet planet. And this is a very important shift in viewpoint. We know, from spectroscopic evidence, that isolated stars (the 15 percent not in multiple systems) invariably rotate slowly, like the Sun. This means that they have given up angular momentum, almost certainly in the formation of planetary systems like our own. We know, from the physics described in the previous chapter, that rocky, "terrestrial" planets must form in the inner parts of such systems, with gas giants further out. And we know, as the near miss of Mars and the presence of life on Earth shows, that the livable band of orbits around a star like the Sun covers a broad enough range to be virtually certain of covering one planet and with a good prospect of covering two.

The pessimists who say that if the Earth were closer to the Sun it would fry, while if it were further away it would freeze, are too gloomy by far. Almost anywhere from halfway to Venus to all the way to Mars our planet would be wet and livable. The air that we breathe and the oceans that surround us are not rare freaks in the

Universe, but the inevitable accompaniments of a rocky planet at such a distance from its parent star that the equilibrium surface temperature, before the atmosphere formed by outgassing, was at least a few degrees above the freezing point of water, and not so high that any water released by outgassing stayed as vapor in the atmosphere. This is a small enough range to make it unlikely that any *particular* planet will be wet, but large enough to ensure that one planet out of several in a system like the Solar System will be wet. The pessimists are being Earth chauvinists—of course it matters to us personally if the Earth lives or dies, but in terms of life in the Universe it could just as easily have been Venus or Mars that won the lottery in our Solar System.

Stars just like our Sun, indeed, may be ideal breeding grounds for life. More massive stars, as we have seen, run through their life cycles very quickly, and life has taken thousands of millions of years to evolve on Earth to a stage where intelligence has arisen. Cooler, fainter stars live longer than the Sun, but their life zones are narrower, decreasing the chance of a planet being in the right orbit to be wet. It may, indeed, be not only possible but likely that life exists on other planets circling other suns; and if it does, then the chances are very good that it lives under blue skies with white clouds and a yellow sun, with rivers and oceans of water nearby in plenty.

THE SHIFTING CRUST

Before we go on to look at just how life got a grip on our own planet, however, there is still one aspect of the changing Earth that needs to be mentioned. The Earth that life developed on, or even the continents that were first trodden by animal life, looked very different from the Earth today in one respect which is, in a sense, superficial, but which would be immediately noticeable if any time-traveling astronaut could bring back a snapshot of the view of Earth from space a thousand million years ago. Although always a wet planet studded with continents, the Earth has not always had the same arrangement of continents and oceans. Continental drift, mountain building, and other geophysical processes have changed the face of the Earth repeatedly, and often with important consequences for life on Earth. Earthquakes, volcanoes, and the ebb and flow of ice ages are all related to these changes, and are all major features that have combined to produce the environment we regard as home today.

Just twenty years ago, the concept of continental drift, in its reincarnation in the form of the modern theory of "plate tectonics," was about to revolutionize our understanding of the Earth. Today, although a very few antidrifters still raise their voices from time to time in scientific circles, the concept is so well established that to the present generation of geographers and geophysicists it seems second nature, common sense in the way that the working of gravity is so obvious once the genius of Isaac Newton has pointed it out to us lesser mortals. In the story of how the idea that the solid continents beneath our feet move around the globe became established, however, it is impossible to point to one genius whose work convinced the scientific establishment. The man widely regarded as the "father" of the concept is Alfred Wegener, a German astronomer turned meteorologist, who published his version of the idea as long ago as 1912 and in detail in 1915. But the convincing evidence only came in the 1960s, through a combination of new techniques for studying the nature of the Earth's crust, especially the crust beneath the oceans, and high-speed electronic computers to analyze the observations. Until the 1950s, indeed, only a few geologists saw the Earth as anything but a very stable place. Continents could rise and fall a little as they responded to the weight of ice added or removed as ice ages came and went, but they could not, it was thought, shift their positions across the surface of the Earth. Ocean floors were thought to be much older than the rocks of the continents, and the most nearly unchanged since the Earth formed. All of these ideas have now been transformed, except that modern geophysicists do still see evidence that the ebb and flow of ice sheets causes vertical displacements of the crust.

We now know that the ocean floors are the youngest features of the crust, and that all of the thin crust beneath the seas is being recycled continuously back into the Earth's interior, so that the oldest rock of the seabed is no more than 200 million years old, compared with the 4,500 million-year age of the planet. The continents not only contain the oldest rocks, but have also been moving about the surface of the Earth, carried by the same processes which recycle seafloor crust, for a very long time. And it is this combined understanding of the behavior of the continents and the recycling of the crust beneath the sea that makes the modern theory of plate tectonics so much stronger than earlier theories of continental drift.

According to legend, one of the first people to notice the similarity in the shape of the eastern coastline of South America and the western coastline of Africa was Francis Bacon in 1620. Bacon does not seem to have suggested that the reason why the two continents fit together like parts of a jigsaw puzzle is that they were once joined and have since drifted apart; but this jigsaw puzzle "fit" has been a source of inspiration to proponents of continental drift down the centuries since. In 1858, the American Antonio Snider developed a version of the concept which also brought in the Biblical flood, suggesting that an original single land mass had split apart, with the waters of the flood rushing into the crack between the Old World and the New; other nineteenth-century scientists toyed with versions of the idea, including theories involving the Moon being ripped out of the Pacific basin, but Wegener's version of the early twentieth century was the first really complete theory of its kind, drawing on evidence from many different scientific disciplines to build up a strong case.

The case was not accepted at the time, nor for another half century, and one reason for this seems to have been that Wegener was an outsider. An astronomer turned meteorologist was not, in the eyes of the geological establishment of the time, someone to be taken seriously, and his position as an outsider helped to ensure that the scientific argument about continental drift was often acrimonious over the next few decades. All that, however, is now behind us, and with the evidence that Wegener gathered plus modern evidence, the idea of continental drift is as well founded as anything in science today.

The jigsaw puzzle fit of the continents, including South America and Africa, is as impressive as ever, but it is actually one of the lesser pieces of evidence in favor of continental drift. When reconstructing the picture of how the present continents can be

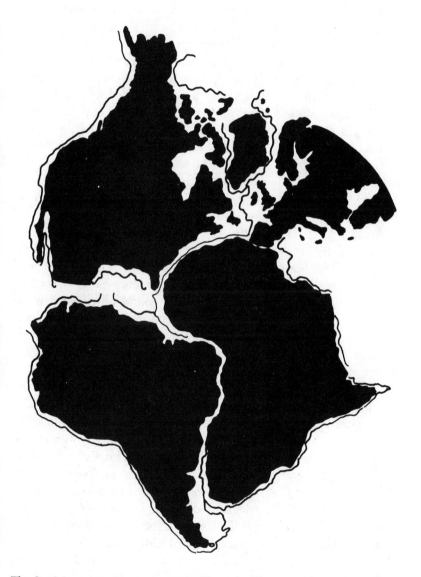

The fit of the continents on either side of the Atlantic, joined not at the present-day coastlines but at the edges of the continental shelf, provides one of the most visually striking indications of the way continents have drifted.

joined together into one supercontinent, geophysicists now use the detailed shapes not of the continents outlined at sea level, but the outline of the continental shelf, which marks the boundary between the thick crust of the continent and the thin crust of the seafloor. This is the natural boundary to choose, but, of course, could only be chosen once it had been recognized, after the modern measurements which showed how thin the oceanic crust really is. With the proper continental boundaries defined (and using modern computers to speed up the task of finding the "best" fit of the pieces of the puzzle, although the job can be done by hand using cutouts slid over the surface of a globe), the fit between South America and Africa is found to be almost perfect, with the other continents fitting neatly together into two supercontinents, dubbed "Laurasia" in the Northern Hemisphere and "Gondwanaland" in the Southern Hemisphere.

A second line of attack backs up the evidence that the present-day continents were joined in this way as recently as 180 million years ago. Just fitting the pieces together does not prove that they "belong" together—in legal terms, the evidence is only circumstantial. But when geological features are found which overlap the join between two continents in the reconstruction, then the evidence becomes far more solid. If we imagine that the jigsaw puzzle pieces are cutouts made from a newspaper, and that these have been juggled about to give the best geometric fit, it is as if we then found that the lines of newsprint on, say, "South America," joined up precisely with the lines of newsprint on "Africa," so that each line could be read across the join, while the two pieces together made up a whole page of a story. Many such "lines" of geological evidence can be "read" across the joins in the supercontinents of Laurasia and Gondwanaland, and only a few examples need be highlighted here. First, old rocks on one side of the join lie next to old rocks on the other side; secondly, the pattern of erosion by ice sheets, wind, and water is the same for rocks on both sides of the join, showing that, for example, southeast Brazil and southwest Africa were joined between about 550 million and 100 million years ago, and then gradually separated*; and as a final clincher,

* The effects of erosion at different times in the past are determined by looking at different layers of rock and sediment. By and large the newest deposits are at the surface of the

continental drift explains why remains of coral reefs (formed in tropical water) and coal deposits (remains of tropical jungles) are found in high latitudes such as Europe.*

Wegener's theory envisaged the continents moving through the weaker crust of the ocean floor, rather like icebergs moving through the water of the sea. Although the modern version of plate tectonics sees both oceanic and continental crust moving together, except at the edges of the plates themselves, Wegener's theory did require two different kinds of crust around the Earth, a thin skin forming the ocean floor and a much thicker, more stable crust making up the continents. And in this respect, while not being completely "right," he was certainly pointing in the right direction. It is far from easy to imagine the "solid" crust of the ocean floor being able to part to make way for the passage of continents, but there are many substances which show the right kind of plastic behavior to allow this kind of deformation. Rock itself, under the right conditions, certainly does deform plastically, as the presence of great mountain ranges and lesser wrinkles in the Earth's skin shows; in more homely terms, the "silly putty" which children of all ages love is another example. This is a rubbery substance which can be rolled into a ball with an extraordinary degree of bounce, stretched into a thin sheet and, if left on a slope, made to flow very slowly downhill. But if the ball is hit suddenly with a hammer, it shatters like a piece of glass. "Solid" rock, subjected to steady pressure over millions of years, can deform plastically, and only appears constant and unchanging to us because of the short span of a human lifetime compared with the life of a continent or mountain range. The shakes and shudders of volcanic and earthquake activity are the hints, dramatic enough in human terms, of these slow processes which are constantly changing the face of the Earth and rearranging the pattern of the continents.

---

ground, with successively older layers beneath. Even though this simple arrangement gets distorted as rocks buckle under the forces involved in continental drift, strata corresponding to different epochs can still be identified and dated accurately using various techniques including the radioactive decay "clock" of certain isotopes.

* Redistribution of the continents by drift also explains many other features of the distribution of fossil remains in the rocks of different parts of the world—the "fossil record." More of this shortly, when I wish to look at the effects of the changing distribution of the continents on life.

During the 1950s, seismic studies showed that the Earth's crust really is thinner under the oceans than the continental crust, averaging only 5 to 7 km in thickness compared with the average 34-km thickness of continental crust (and in places 80- or 90-km thickness beneath great mountain ranges). This did not, however, confirm that Wegener's idea of continents moving *through* the oceanic crust was correct. The same survey techniques which revealed the thinness of the oceanic crust also showed that the surface of the crust beneath the sea is as ruggedly sculptured in its own way as the surface of any land mass, with submarine mountain ranges and canyons, very deep trenches in some parts of the world and, most significantly of all, a great ridge system which runs around the world, with branches extending into every ocean. The branch of this ridge system which runs up into the North Atlantic was soon studied in great detail, and has since become accepted as the archetype of oceanic ridges. It rises about 3 km above the plains of the seafloor, and it is a very active geological region, where submerged volcanic activity and other subterranean rumblings are common—the activity is most easily seen in Iceland, where the ridge actually breaks surface in the form of that volcanic, geologically active little island.

CONTINENTAL DRIFT AND SEAFLOOR SPREADING

In 1960, this ridge system and its activity were explained by Professor Harry Hess of Princeton University in terms of seafloor spreading, reviving the idea of continental drift and putting it on a firm basis at last. According to the new version of this old idea, material from the inside of the Earth wells up through convection in some parts of the world, cracking the crust above and pushing new, molten rock out of the cracks, which grow to become the ocean ridge system. The thin crust produced by the solidifying rock is continuously pushed outward from the ridge, spreading out on either side to make up a growing ocean floor. This idea explains very neatly the fit between the edges of continental shelf on either side of the Atlantic, as the cracked-apart pieces of a larger, older continent, which have been pushed steadily apart for nearly 200 million years as the activity of the Atlantic spreading ridge has steadily widened the ocean between them. In the case of the Atlantic, the widening occurs at a rate of about 2 cm per year (spreading at a rate of 1 cm per year on each flank of the central ridge).

This whole process is seen as very much only "skin deep" in terms of the Earth as a whole. The mobile crust, cracked and pushed about by the convection beneath, is simply the lithosphere, a skin no thicker, compared with the whole Earth, than the thickness of the skin of an apple compared with the whole apple. And the fluid region beneath the lithosphere, in which convection currents do the stirring, is only the partially molten asthenosphere, 75 to 250 km below the surface itself. All of the activity involved in continental drift takes place within the top 250 km or so of the Earth's surface, driven by the heat escaping from the interior, which stirs the convection currents. Out of a radius of 6,371 km, this represents merely the outer 4 percent of our planet. If this spreading was the only important process involved in continental drift, then clearly our planet would have to be getting bigger to accommodate all the new crust. But at the same time as the mapping of the great ocean ridge system was taking place, geophysicists were learning about the existence of deep ocean trenches, which edge the western side of the Pacific from New Zealand through the Philippines and Japan to the Aleutians in the north, and the eastern side of the Pacific from Chile to Mexico. Hess suggested that the new ocean crust being created at spreading ridges was exactly balanced by the destruction of relatively old ocean crust at these deep trenches, where the downward limb of a convection cell sweeps the relatively thin oceanic crust down under the much thicker crust of a continental mass back into the asthenosphere, where it melts, and from which the material may be recycled through another—or the same—branch of the spreading ridge system.*

---

* I should mention, however, that there are respectable geologists to be found in the world today who argue that the Earth has indeed expanded somewhat since it first formed as a solid planet, while another vociferous minority argues equally strongly that our planet has shrunk a little over geological time. But the consensus is that no dramatic changes in size are occurring today, and certainly the kind of large-scale expansion that would be required to produce the Atlantic Ocean in only 200 million years is not regarded as a serious theory. Apart from anything else, such rapid expansion would have changed the radius of the Earth so much over the past 200 million years that the "jigsaw puzzle" pieces of the continents could no longer be fitted together on the globe, having broken apart when the Earth's cur-

This very simple model immediately slotted the pieces of the new world picture into place. The continuing volcanic activity and outgassing of the ocean ridge system helped to explain where all the water of the oceans had come from; while the idea of an active ocean crust, being continually recycled, and a more enduring continental crust gave a new perspective on the geological activity of the Earth, explaining why the main action in geological terms takes place where it does. And the clinching evidence which confirmed the accuracy of the seafloor spreading hypothesis beyond all reasonable doubt came early in the 1960s, as geophysicists began to measure and study the patterns of magnetism "frozen in" to the rocks of the seafloor.

The rocks of the seafloor are mainly basalts, which are weakly magnetic. When molten basalt squeezes out from an oceanic ridge and spreads on either side, it cools quickly and solidifies as it encounters seawater, and as it sets it has imprinted within it a magnetic field aligned with the magnetic field of the Earth. Once the rock has set, however, in magnetic terms it becomes rather like a solid bar magnet, and if a piece is broken off and moved around, or moved in some other way, it carries the same magnetic field oriented with the piece of rock, not with the Earth's magnetic field. Where layers of rock have been laid down one on top of the other by repeated volcanic activity on land, this effect can be clearly seen in successive layers of rock, because the Earth's magnetic field is not constant. The orientation of the magnetic poles shifts relative to the geographic poles, and in addition, for reasons which are still far from being well understood, the entire field "flips" from time to time, with north and south magnetic poles swapping places. So successive layers of basalt on a continent may contain a surface layer aligned with the Earth's present magnetic field, a layer or layers aligned slightly differently, matching the position of the wandering magnetic poles in the recent past, and layers with a completely reversed magnetic polarity, corresponding

---

vature was much less. Any debate about long-term and subtle changes in the Earth's size (or dramatic but long-ago changes) has no bearing on the evidence that at present—by which geologists mean over the past few hundred million years—the size of the Earth is essentially constant and that drift has taken place on a globe which stays the same size.

to times in the geological past when the Earth's field was opposite from the orientation it has today.

When magnetic observation of the seafloor in different parts of the world showed stripes of magnetic patterns running roughly north-south, but with adjacent stripes having opposite magnetic polarity, it didn't take long for the geophysicists in the middle of the 1960s to come up with an explanation, based on the seafloor spreading idea. Quite simply as basalt spreads out from an ocean ridge today, it sets with the imprint of the present-day magnetic field. When the field reverses, new basalt will now set with the opposite polarity, alongside the older basalt, which is still being carried steadily away from the spreading ridge. As this process repeats itself many times over a hundred million years or more, a pattern of magnetic stripes builds up, distributed symmetrically around the spreading ridge, in which each magnetic stripe can be dated by comparison with the patterns of magnetic reversals found in rocks of different ages on land.

The concept is brilliantly simple, although the observations of this magnetic "tape recorder" are far from being as straightforward as I have made them sound. The result is clear-cut: the seafloor spreading idea really does explain what is happening to oceanic crust, new crust really is created at ocean ridges and really does disappear down the throats of ocean trenches, forming a kind of global conveyor belt system, on which the continental crust is taken along for the ride. Between spreading ridge and ocean trench lies a region of relative stability, a "plate" of the Earth's crust. Such plates can also rub shoulders, with crust being neither created nor destroyed at the boundary, giving three types of "plate margin": where new crust is created we have a constructive margin, where old crust disappears we have a destructive margin, and where plates rub shoulders we have conservative margins and transform faults like the San Andreas in California, which separates the North American plate from the Pacific plate.

Magnetic reversals happen very quickly in terms of the Earth's lifetime, which makes for good, sharp boundaries between the magnetic stripes. The whole process takes a few thousand years, and once the magnetic field has become established in one direc-

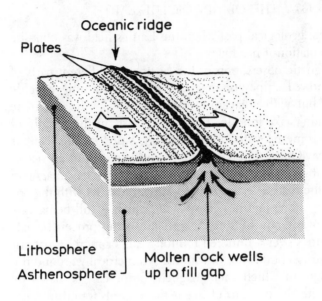

Oceanic ridge

Plates

Lithosphere
Asthenosphere

Molten rock wells
up to fill gap

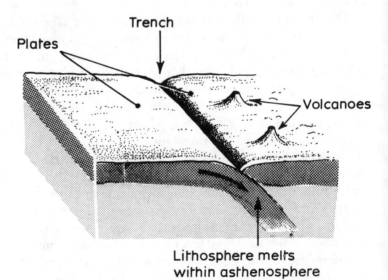

Trench

Plates

Volcanoes

Lithosphere melts
within asthenosphere

The mechanism of continental drift is seafloor spreading. At ocean ridges, cracks
in the thin crust of the seafloor, new rock is laid down as molten material wells
up from beneath and spreads on either side of the ridge. The widening of the
seafloor that results is exactly balanced where thin oceanic crust is forced down
under the thicker crust of a continental mass, forming a deep trench at the edge
of the continent and buckling the continent to build mountains and volcanoes.

tion or the other, it may stay that way for as little as a hundred thousand years or for as long as 50 million years. At present, the Earth's magnetic field is slowly weakening, at a rate which would make it disappear in about 2,000 years' time, so it is just possible that we are living through the early stages of the "next" magnetic reversal. One side effect of this magnetic activity may be relevant to the story of life on Earth. It has been suggested that when the magnetic field is weak or absent, then the increased flood of charged particles from space—cosmic rays—which reach the surface of the Earth could be damaging for life, especially land-based life. Normally these charged particles are deflected by the Earth's magnetic field, stored in the Van Allen radiation belt, and spill over near the magnetic poles to produce the bright aurorae seen at high latitudes. If, without the protecting magnetic field, they reached the ground or the lower atmosphere directly, they might prove directly harmful to life, or they might disrupt the climate, which in turn would make life on land difficult. Such ideas are still speculative, although some speculators do point out that the epoch which saw the "death of the dinosaurs," some 65 million years ago, coincided (if that is the right word!) with a magnetic reversal. Even though this aspect of the Earth's history is speculative, however, it is clear that the movements of the continents through the processes of plate tectonics have been crucial to the evolution of life on our planet.

At present, the Earth's crust can be divided up into six large plates, seismically quiet zones which may contain all oceanic crust (like the Pacific plate), predominantly crust (like the Eurasian plate), or a mixture of the two (like the North American plate, which stretches from the east Pacific seaboard to the mid-Atlantic ridge). In addition, there are about a dozen smaller subplates, plus some complex jumbled regions at the boundaries, especially where three or more plates meet together, as in the Caribbean. Great mountain ranges are produced where plates collide—around the Pacific Ocean, where the activity has produced a belt of volcanism which was known as the "Ring of Fire" long before the idea of plate tectonics became established; in the Himalayas, where India is colliding with Eurasia; and in the Alps and Pyrenees, where Africa is colliding with Europe. Where the collision brings conti-

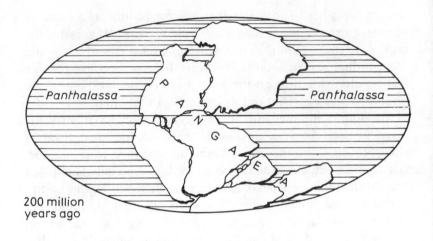

200 million
years ago

135 million
years ago

The breakup of Pangaea II is the most recent major event in the Earth's tectonic history, and is now well understood from the record in the rocks. A key feature, as far as life on Earth is concerned, is the way the continents are grouped today to ensure that both poles can be ice-covered at the same time. This rare geographical arrangement makes the Earth prone to Ice Ages, and allows the astronomical

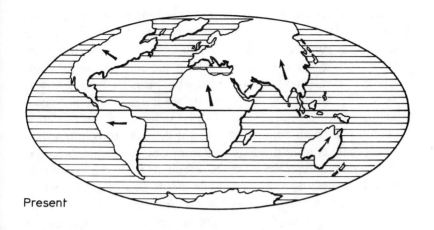

Present

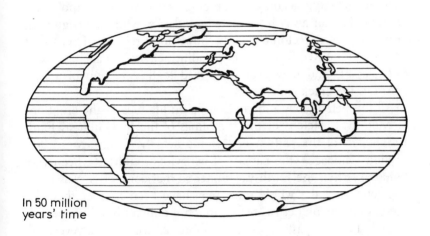

In 50 million
years' time

cycles of the Milankovich models to be effective. Continental drift has not, of course, halted today, and the North Atlantic, for example, is widening at a rate of about 2 cm per year. In 50 million years from now, the geography of the world will be noticeably different, and the easier passage of warm ocean currents into the Arctic may end the present cycle of Ice Ages.

nental crust against continental crust, sediments which have accumulated on the continental shelf are squeezed by the pressure of the closing jaws and molded into the rocks of new mountain regions, being squeezed upward like toothpaste from a tube and helping to create new continental crust. This is happening now in the Mediterranean region, where before long, on the geological timescale, all the seafloor crust will be squeezed out of existence. When a plate with continental crust at the edge runs into a plate with oceanic crust at the edge, the thin oceanic crust is forced beneath the continental crust while sediments from the seafloor are scraped up, helping to build coastal mountain ranges. Meanwhile, the ocean crust below is forced downward, breaks off, and melts, with the hot material rising and producing volcanic activity, while the continental material itself buckles under the strain. Nowhere is this process more clearly seen than in the mountains which form the spine of South America, down the western edge of the South American Plate. Such processes long ago, in previous cycles of tectonic activity, account for the presence of older mountain ranges today, far away from plate boundaries, while the beginning of new phases of tectonic activity shows up in regions such as East Africa and the Red Sea, where spreading activity is cracking a continent apart, and may eventually lead to the creation of a new ocean as the crack widens.

All this activity has a direct bearing on life today—the eruptions of Mount St. Helens or the rumblings of the San Andreas Fault are all part of the pattern of tectonic activity. Long ago—3,000 million years ago—the pattern of behavior would have been very different, when the Earth was hotter so that convection was more energetic, and the whole crust may have been thinner than the present continental crust, before much mountain building had occurred. For the past thousand million years or so, however, the processes of plate tectonics have probably occurred much as they do today, rearranging the distribution of continents and oceans in four major phases of activity. And these rearrangements seem to have had an effect on the life of the times, an effect so important that it seems relevant to mention it here, just before we turn our attention to the full story of life on Earth.

In geological terms, everything that happened more than 570

million years ago (about 90 percent of Earth history) is lumped together as the Precambrian, and very little is known about it. As we come closer up-to-date, the divisions of geological time become shorter, corresponding to the much greater knowledge we are able to glean from the geological record regarding more recent times. During the 570 million years or so for which we think we have a fair idea of what was going on on Earth, the continents seem to have first broken up from one supercontinent to form four separate continents, then got back together again in a new supercontinent, and finally broken up again, first into Laurasia and Gondwanaland and then into the seven continents we know today. This pattern of events coincides intriguingly with the changing pattern of life on Earth, and in particular with the number of different species found in the fossil record for each epoch, especially the fossil remains of creatures that lived in shallow seas.

To explain this, imagine the situation when the Earth contained one supercontinent ("Pangaea"). Such a great land mass would have very hot summers and very cold winters, since the moderating effect of the oceans could not reach far inland. It would also have one offshore marine "environment" stretching around its coastline, so that sea creatures could spread everywhere around the coast. Now imagine what happens when the continent breaks up and separates into a variety of smaller continents. First, the climate is likely to be less extreme (except in special cases, of which more later), since the maritime influence can reach a bigger proportion of the total land cover of the planet. Secondly, each continental shelf would be cut off from the others, so that sea creatures which develop along one coastline cannot spread out over the entire coastline of the globe. Both conditions will encourage a greater diversity of life—less extreme climate allows diversity at each coastline, and the separation of coastlines allows different forms to develop independently, without a few dominant forms wiping out all opposition. So a collection of smaller continents should produce a greater variety of fossil remains in the record of the rocks than one single supercontinent. And that is exactly what we find.

As Pangaea I broke up some 600 million years ago, faunal diversity increased dramatically, until the four new continents were established by about 450 million years ago, at the end of the

Ordovician. This diversity continued until about 200 million years ago, when the four continents converged with one another during the Permian and formed another supercontinent, Pangaea II. At this time, the number of different species found in the fossil record drops dramatically; but Pangaea II was short-lived, quickly breaking up again into a new arrangement of continents, first Laurasia and Gondwanaland separated by a new ocean, Tethys, and then the further breakup which began to give an appearance rather like the present-day Earth, but with the continents in slightly different positions, some 70 million years ago, at the end of the Cretaceous.* The diversity of species on Earth increased throughout this phase of breakup, first from one supercontinent into two, then from two into seven, with the world today providing a variety of interesting homes for dwellers of the continental shelf, more diverse than at any time in the past 600 million years. If the continents had never moved about and been regrouped in this way, then the evolution of life on Earth would have been a lot simpler, a lot duller, and, quite possibly, might never have resulted in the arrival of an intelligent species.

This, clearly, is my cue to bring the story of life, at long last, into the picture, even though my discussion of the 1960's "revolution in the Earth sciences" has necessarily been so superficial in the space available. There is, however, one last piece of physical background to sketch in, a secondary revolution in the Earth sciences which came in the 1970s, and directly followed from the improved understanding of the solid Earth built up in the previous decade. This second revolution concerns a topic every bit as relevant to the story of life on Earth, and particularly relevant to the story of human life—the causes of climatic change, and the ebb and flow of great ice sheets across our planet.

OUR CHANGING CLIMATE

Geologists can reconstruct the history of past climatic changes from the way these have affected sediments being laid down at different times. In the most extreme example, the movements of ice

---

* Speculators might like to ponder on whether this event, rather than a magnetic reversal, might have had significance for the life and death of the dinosaurs!

sheets across the rocks during ice ages gouges out scars which remain visible for hundreds of millions of years after the ice has retreated. By dating different rock strata, using a variety of techniques and looking for traces of the passage of ice, geologists can confidently say when (give or take a few million years!) the ice covered a particular piece of continental crust. But that is only half the story; they also need to determine *where* that particular piece of the Earth's crust was when the ice covered it, since we now know that the geography of the globe is constantly changing as the continents drift—or, rather, as they are pushed about by the mechanism of seafloor spreading on the global conveyor belt. The reconstructions of position depend partly on matching up geological patterns on different continents today, the "lines of newsprint" already mentioned, and are aided by matching up the fossil magnetism of rock strata in different parts of the world. Continental crust, too, contains rocks with frozen-in magnetism, showing the orientation and, to some extent, the location of a particular piece of continent at the time particular strata were being laid down. So, one way or another, it is possible to build up a reliable guide to the broad sweep of climatic changes over the past thousand million years or so—the ebb and flow of ice ages occurring at rare intervals during much longer epochs of a warm, wet, "tropical" world. And this can be compared with the geography of the world—the distribution of continents—in some detail back to about 600 million years ago, and rather more vaguely before then.

On this sort of timescale, the "normal" climate of the Earth is seen to be warm and wet. At intervals of something like 250 to 300 million years, a change occurs which brings a colder climatic regime, an ice "epoch" which may last for 50 or 100 million years. Within such an ice epoch, the glaciers may ebb and flow many times, sometimes covering a great deal of the land surface of the Earth, and at others retreating to smaller ice caps around the poles. Such intervals between the full ice ages of an ice epoch are called interglacials, and we are living in an interglacial today.

There have been many speculations about the causes of these rhythms of climatic change. On the longer scale, the recurrence of ice epochs on a usually warm world has been blamed by some on changes in the Sun itself and by others on the passage of the Solar

System through clouds of dust in our Milky Way Galaxy. (Remember that a couple of hundred million years is, very roughly, the interval between successive passages of the Solar System through the dust lanes which edge the spiral arms of the Milky Way.) Both ideas are possible, though not necessarily plausible. But the best suggestion so far relates these long-term changes in climate to the long-term changes in the geography of our planet.

It is natural to start out by looking at the present-day situation, which we know most about, and then to compare this with the changes that have taken place in the past few hundred million years. Today it is rather obvious that the coldest, most icy places on Earth are at the poles, where the Sun appears low on the horizon and the least solar heat reaches each square meter of ground. The atmosphere and the oceans of the world compensate for the imbalance in solar heating to some extent, as atmospheric convection and, most importantly, oceanic circulation, carry warm air and warm water out of the equatorial belt between the tropics and up to the high latitudes. With the present geography of the world, however, the warm water cannot penetrate all the way to the poles. In the south, the continent of Antarctica sits squarely over the pole, so no water at all can get to the pole itself, except the water which falls from the sky in the form of snow, and has built up the greatest sheets of ice found on Earth today. In the north, although there is a polar sea, it is almost entirely surrounded by land masses, so that the great warm currents, such as the Gulf Stream, cannot penetrate in full strength, and a skin of ice—thin compared with the South Polar ice cap, but still important— remains over the Arctic sea the whole year round. So the present ice epoch, a series of ice ages which began about 10 million years ago, can be very well understood in terms of the arrangements of the continental land masses today. When we find, as we do, the remains of tropical forests in Antarctica (in the form of coal deposits), this does not mean that the South Pole was once tropical, but that Antarctica was once part of a land mass in the tropical regions; when we find traces of glaciation in old rocks in Brazil, we do not assume that once, long ago, the tropical regions of the Earth were covered by ice, but rather that Brazil used to be part of a polar continent.

Whatever else is going on in the world, glaciers of some kind can form whenever any continental mass drifts over one of the poles, blocking out the warm water and building up a layer of snow which develops into an ice sheet. Sometimes this must happen when the other pole is still open to warm water and the other continents are arranged in low latitudes near the equator. But full ice epochs, like the present one, depend on the simultaneous presence of land at both poles—either actually over the pole, as with Antarctica today, or surrounding the polar sea, as in the north today.

A great ice epoch which was centered on the period around 700 million years ago, at the time of Pangaea I, may well have been one of the first kind. Part of the supercontinent overlapped one pole, allowing ice to form. But with very little land apart from the supercontinent, the other pole must have stayed warm and wet. When Pangaea I broke up and its four main pieces wandered their separate ways, warm water could get to both poles again and the Earth stayed warm for some 300 million years, while both flora and fauna flourished and diversified. Then came the reassembly of the separate continents into Pangaea II, the supercontinent which preceded the present phase of the Earth's tectonic activity.

It is from the time of Pangaea II that we really have a very good idea of how seafloor spreading and continental drift have shaped the changing face of the Earth. During the assembly of Pangaea II, a land mass made up of present-day South America, Africa, Antarctica, and Australia drifted across the South Pole and resulting glaciation left its scars. Even after Pangaea II was assembled, rather less than 300 million years ago, its southern tip extended across the South Pole and ensured continued glaciatian, although a great deal of the land mass stretched up across the tropics. Again, however, as in the time of Pangaea I, the presence of just one great land mass meant that only one pole could possibly be ice covered, and the North Polar sea remained warm, fed by currents of warm water from the tropics. By about 250 million years ago, after some 50 million years of this one-sided ice epoch, Pangaea II drifted away from the South Pole and began to break up. For 200 million years, the Earth as a whole was again warm, with snow and ice simply local phenomena found on high mountains. Once again, the fauna and flora spread and diversified, until about 55 million years ago the world as a whole started to cool. Antarctica, now split off from Australia, moved slowly into its present position over the South Pole; the northern continents edged into their present positions around the North Pole as the Atlantic widened and the great ocean on the other side of the world began to be squeezed out of existence, developing as the Pacific Ocean we know today. About 10 million years ago, the present ice epoch was fully established, with glaciers appearing in Alaska and other northern regions, and the Antarctic ice sheet building to about half its present size. By 5 million years ago, the Antarctic ice sheet may have been even bigger than it is today, and some 3 million years ago great ice sheets made their first appearance—in the present ice epoch—on the continental masses of the north, alongside the North Atlantic Ocean. Once this pattern had been established, the ice sheets began a long and complex pattern of advances and retreats, following three main rhythms roughly 100,000, 42,000, and 22,000 years long. These rhythms are behind the ebb and flow of ice ages and the occurrence of interglacials. Mankind has arisen during the present ice epoch and our civilization has arisen during the present interglacial, so these rhythms are of more than just passing interest. And it is these rhythms that were explained satisfactorily in the second revolution in the Earth sciences, in the 1970s.

Before we look in detail at the driving force behind ice ages, so important to human life on Earth, it is worth stressing the unusual nature of present-day geography. Other ice epochs have occurred in the past thousand million years, but apparently only when a single land mass has drifted across one pole, leaving the other pole ice free. Conditions which are able to produce ice-covered poles in both hemispheres at the same time are clearly rather unusual in the history of the Earth. The stresses (as far as life is concerned) that result played a part in encouraging intelligent life to develop on Earth, since when life is easy and the environment is unchanging, specialization, rather than the adaptability provided by intelligence, is more likely to ensure the survival of a species. Here perhaps is the reason why *we* have appeared on Earth at the present moment of geological time, children of the present ice epoch and grandchildren of the present arrangement of the continents, products, in a very real sense, of the processes of plate tectonics.

THE RHYTHM OF ICE AGES

The saga of the developing understanding of the rhythms of ice ages within the present ice epoch has curious echoes of the unfolding story of the understanding of continental drift. The theory, which explains the ebb and flow of ice ages in terms of changes in the orientation of the Earth with respect to the Sun and changes in the Earth's orbit around the Sun, is generally called the Milankovich model, after Milutin Milankovich of Yugoslavia, who spelled out details of the idea some forty years ago. But just as the idea of continental drift predated its great prophet Alfred Wegener, so the astronomical theory of ice ages predated Milankovich, and the two stories are nicely linked, since one of the supporters of an older version of the concept, pre-Milankovich, was Alfred Wegener himself. Both ideas survived for decades before being finally proved correct, and in both cases the proof rested upon new observations about the Earth and history made possible by improved techniques—neither the idea of continental drift nor the Milankovich model of ice ages could possibly have been proved correct at the time they were first formulated, and both Wegener and Milankovich were, in that sense, ahead of their time with their theories.

Whereas the driving force behind continental drift is still only poorly understood, however, Milankovich started out with a very well-known astronomical driving force and sought to explain the observed pattern of ice ages within its framework. Three separate cyclic changes in the Earth's movements through space combine to produce overall changes in the pattern of solar radiation arriving at different latitudes of the Earth at different times of the year, and these changes are the key to the model. The longest, but not necessarily the most important, is a variation which repeats itself over a period of between 90,000 and 100,000 years, during which the shape of the Earth's orbit around the Sun stretches slightly from being almost circular to being slightly elliptical, and back again. This does not change the total amount of heat arriving at the

Earth from the Sun; the average distance of the Earth from the Sun* is always the same. But when the orbit is more nearly circular, there is a more nearly even spread of the heat over the whole year, and when the orbit is more elliptical, then, since the Earth is closer to the Sun in one part of its orbit, some months are warmer than others. This can increase the contrast between the seasons, even though the average heating over the whole year is unchanged.

Secondly, there is a cycle roughly 40,000 years long over which the tilt of the Earth changes. At present, the Earth is tilted by 23.4° out of the vertical, compared with the plane of its orbit around the Sun, or with a line joining the center of the Sun to the center of the Earth. In July the North Pole points toward the Sun, and we have summer in the Northern Hemisphere. In January, when the Earth is on the other side of the Sun, the North Pole still points the same way through space, but this direction is now away from the Sun, and we have Northern Hemisphere winter. The reverse, of course, is happening in the south, giving the pattern of seasons we know so well. The tilt of the Earth varies between 21.8° and 24.4° over the 40,000-odd–year cycle already mentioned, and when the tilt is more pronounced there is more contrast between the seasons, while millennia with less tilt have a more even distribution of solar warmth over the Earth throughout the year.

Finally, although the orientation of the Earth in space does not change significantly in the course of one year, the Earth is actually wobbling like a spinning top as it gyrates through space, with the direction in which the polar axis "points" following a circle around the sky with a period of 21,000 years. This wobble also changes the amount of heat reaching different parts of the Earth in different seasons, although, again, it has no effect on the total amount of solar heat ("insolation") arriving at the Earth in the course of a whole year.†

So the three effects combine to produce a seemingly complex

---

* Arranged over a year or many years.
† I have glossed over how we know about these changes. In fact, it is straightforward (if tedious) to calculate the various gyrations of the spinning Earth from the gravitational effects on our planet of the Sun, Moon, and other planets of the Solar System, notably Jupiter; in

pattern of variations in seasonal heating of the Northern and Southern Hemispheres, while leaving total insolation unchanged. And it is easy to see how such seasonal changes in insolation could cause the ebb and flow of ice ages—given the present, rather peculiar, distribution of the continents of the Earth.

THE SITUATION TODAY

Looking at the Earth today, what changes in seasonal heating would be required to encourage the spread of ice at high latitudes? In the Northern Hemisphere, where there is plenty of land around the cold polar sea, any snow which falls can lie on the ground and relatively little snow falls into the ocean and melts. Winter can be relied on to bring some snow somewhere over the northern continents. So the best way to make an ice age is to have cold *summers*, so that the winter snows stay lying around for most of the year. Once snow and ice fields become established, they can enhance the cooling effect because they shine so brightly—they have a high albedo—and reflect away a great deal of the incoming solar heat. So according to the Milankovich model, ice ages should develop in the north when the interlocked rhythms of the Earth's gyrations bring the coldest summers.

In the south, however, the pattern is completely reversed. There, the pole itself is covered by a continent and a great layer of ice, and this in turn is surrounded by sea. Snow falling into the sea melts, and the only way to make the ice sheets spread out further is to have winters so cold that more of the sea itself freezes. Once the sea is frozen, even the heat of a warm summer can be reflected away by the spreading ice sheet without all of it melting. To make ice ages in the south, then, we need very severe winters. And as a very little thought shows, the changing patterns of the Milankovich cycles must produce cool northern summers at the same time as severe southern winters, since northern summer occurs at the same part of the Earth's orbit as southern winter!

So far so good—and relatively simple and straightforward. To

---

addition, the shorter cycles at least have produced noticeable changes in the appearance of the stars of the night sky during recorded history, so that the effects can be confirmed directly from observations.

test the theory it was first necessary to match up the patterns of past ice ages with the Milankovich rhythms, then to find which seasons are the most critical for insolation changes, and finally to check that the amount of variation in insolation really is sufficient to account for the volume of ice frozen during an ice age and melted during an interglacial. The first step was where the Milankovich model fell down forty years ago. At that time geologists thought that there had been only four or five full ice ages in the recent past, and the patterns of advance and retreat of the ice they had worked out from the scars left on the rocks of the geological record simply did not match up with the Milankovich model, which "predicted" the occurrence of many more ice ages, separated by short interglacials. So the theory remained in limbo through the 1930s, 1940s, and into the 1950s. But then improving techniques began to change our understanding of the rhythms of past ice ages—and every change brought the picture of Earth history more in line with the calculations made by Milankovich.

The new understanding of the ebb and flow of ice ages depends both on a better supply of material from which information about past climates can be extracted and on improved techniques for extracting the information. The raw material comes from drilling in the ocean floor, which produces cores of sediment that provide a layer-by-layer record of the debris being deposited each year, and from improved studies of similar cores on land, together with cores drilled through ice sheets at high latitudes, where each year a new layer of snow is added to the ice sheet, carrying with it information about the temperature at the time. "Conventional" techniques, such as examining fossil remains to determine whether creatures that loved warmth lived at a certain epoch or whether they had been replaced by more hardy species, played a full part in developing the new picture of climatic changes, but the key technique in the revolution depends on analyzing the isotopes of different elements found at different depths in the cores.

A natural "thermometer" is provided by the proportion of the heavy isotope oxygen-18 in water, ice, and the atmosphere at any time. Because this isotope is heavier than the more common oxygen-16, water molecules that contain oxygen-18 (H_2O-18) are less easily evaporated, but more easily condensed, than molecules containing O-16. The relative amounts of O-18 and O-16 found in different layers of sediment or in different layers of ice in a glacier provide a very precise measurement of the overall temperature of the globe when they were being laid down.*

Putting all of the pieces together, with thousands of seafloor cores and dozens of ice cores accumulated over two decades, the time was ripe in the mid-1970s for a complete rehabilitation of the Milankovich model. By then it was clear that there had indeed been many ice ages, separated by short interglacials, in the past couple of million years, and Jim Hays and colleagues from the Lamont-Doherty Geological Observatory in New York published what is seen by many in retrospect as the clinching evidence in 1976. Their results were based on detailed analysis of two overlapping cores drilled from the floor of the southern ocean, which between them covered a span of 450,000 years, sufficient to show the presence of even the longest Milankovich cycle. The key analysis took samples every 10 cm along the cores, corresponding to intervals of 3,000 years, short enough to test also for the shortest of the Milankovich cycles. And precise statistical tests of the temperature variations revealed by the samples left no doubt in the minds of the Lamont-Doherty team:

> There can be no doubt that a spectral peak centered near a 100,000-year cycle is a major feature of the climatic record. . . . Dominant cycles . . . range from 42,000 to 43,000 years [and] three peaks . . . correspond to cycles 24,000 years long.

The shorter of these periods seem to produce the most dramatic effects, with the long 100,000-year cycle modulating the other influences, so that at times it offsets their effects and at others it adds to them. The statistics also show a trace of a 19,000-year cycle,

---

* The oxygen in the sediments is mainly in the form of carbonates, the chalk from the shells of long-dead sea creatures. The carbonates also provide a means to "date" sediments over about 100,000 years, from the proportion of radiocarbon (carbon-14) which they contain. All of these techniques are difficult and require superb laboratory skills, which is why it has taken so long for the new understanding of ice ages to emerge. The evidence is, however, unambiguous and now completely incontrovertible; credit for initiating the revolution belongs to Cesare Emiliani's pioneering work in the 1950s.

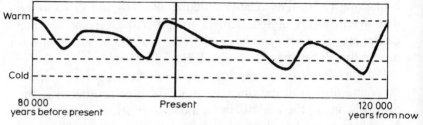

Warm

Cold

80 000
years before present

Present

120 000
years from now

With the present arrangement of continents, the combined rhythms of the Milan-kovich cycles both explain recent climatic variations and give us a long-range forecast. Although continental drift may break up the pattern that allows Ice Ages in 50 million years or so, for the next 60,000 years or more the outlook is decidedly unfavorable.

which is explained by the most modern version of the Milankovich model, updated with the aid of high-speed electronic computers. And the success of the combination of theory and measurements in explaining the pattern of past ice ages gives a clear guide to the future—we should expect a long-term trend over the next few thousand years toward extensive Northern Hemisphere glaciation, with the present interglacial almost over.

What of the other details of the model still to be filled in? By one of the great coincidences which happen from time to time in science, both the other main aspects of the theory were improved dramatically over the same couple of years, 1975–1976, that the best ever seafloor core study was being prepared. Separate cal-culations by Max Suarez and Isaac Held of Princeton Univer-sity, and George Kukla, of Lamont-Doherty, identified the key season for spread of Northern Hemisphere ice as late summer/early autumn, and the key geographical locations as the interiors of North America and Eurasia. And John Mason, Director-General of the U.K. Meteorological Office, provided what can be re-garded as the meteorological "establishment's" seal of approval with calculations which showed that the changes in insolation in the critical seasons over the past 100,000 years closely matched the ebb and flow of ice. We know how much heat is given up when water vapor falls as snow and how much heat is needed to melt each gram of ice back into water. By comparing the "surplus" or "deficiency" on the seasonal insolation budget produced by the changing Milankovich cycles with the volume of new ice being created or melting each millennium over the past 100,000 years (revealed by various pieces of geological evidence concerning the thickness and extent of ice sheets during the most recent ice ages), Mason showed that the amount of ice building up in cold millennia exactly matched the Milankovich insolation deficit, while the amount melting in warm millennia closely matched the Milankovich insolation surplus. With these detailed calculations, he provided perhaps the most accurate "long-range forecast" yet—the escape from ice age conditions which began 18,000 years ago and brought the world into the present interglacial took the combined efforts of all three Milankovich cycles working for warmth; this warmth peaked about 6,000 years ago, when low orbital eccentricity plus the shifting Earth wobble combined to boost the heat of northern summers at the same time that the tilt of the Earth reached maximum, putting the Sun high in the summer sky. Since then, however, all three factors have passed through their most beneficial states, in terms of northern summer warmth, and conditions are getting steadily less favorable. With 60,000 years of unfavorable orbital geometry ahead, the next ice age is due imminently—which, on this timescale, means that it could start in 1,000 years' time or perhaps not for 5,000 years, if nature is left to take its course.

And this is where we come in, both literally and figuratively. Our civilization is a product of the present interglacial, and could be the victim of the next ice age. On the other hand, our technology and both deliberate and inadvertent modification of the environment of our planet may be able, by accident or design, to prevent the "next" ice age from ever developing. It is because climatic changes are so important to life on Earth, including human life, that I have dwelt in such detail on these changes which are so minor in terms of the long history of our planet, but which have played a major part in the origin of the Earth as we know it today. In terms of physical changes in our planet and the Universe around us, we have now come completely up-to-date. For most of its 4,500 million-year history, however, our planet has had passengers in the form of living organisms, and to understand our own origins we have to go back to the origins of life on Earth—and even, perhaps, before the Earth formed.

5 THE ORIGIN OF LIFE

To understand the origin of life, we need to know a little chemistry, the scientific discipline that describes how atoms stick together as molecules to form different substances—different chemical compounds. Inside a star, although different elements may be built up by fusing protons and neutrons together, as long as the star remains hot the material remains as a plasma, a sea of positively charged atomic nuclei swimming in a sea of negatively charged electrons. When we were dealing with the evolution of the whole Earth, we were mainly interested in the behavior of matter in large chunks—a whole ocean of water, a continent of rock—where the details of the structure of atoms and molecules matter less than the overall properties of solid, liquid, or gas (although, of course, those properties ultimately depend on the atomic and molecular properties). But life, and especially the origins of life, depends crucially on the behavior of individual atoms, and the atoms of one element in particular.

Under the conditions of temperature and pressure that prevail on the surface of a planet like the Earth, rather than those that apply in the interior of a star like the Sun, the positively charged atomic nuclei and negatively charged electrons get together to form electrically neutral atoms. The simplest hydrogen atom consists of just one proton, making up the nucleus, and one electron surrounding it. The idea of a single electron being able to surround anything, even a single proton, seems bizarre to human minds used to the behavior of particles in the everyday world (how could a golf ball "surround" a cricket ball?), but I use the term deliberately. When atomic structure was first probed, and the differences between nuclei and electrons investigated, the first imaginary "models" of atoms envisaged the electrons in orbit around each nucleus, like the planets orbiting around the Sun. But electrons don't behave like solid particles in the way that a planet or a billiard ball does, and for many purposes electrons can be described as waves, or wave packets. Just as relativity theory reveals that mass and energy are interchangeable, so one of the other great developments in twentieth-century science, quantum mechanics, reveals that solid particles can be thought of as waves, and energetic radiation can be thought of as particles, depending on circumstances. The classic example of this is light, where the argument about whether light is made up of tiny particles (photons) or is a form of wave energy (electromagnetic radiation) went back to the time of Newton. From Newton's time onward, proponents of each theory were able to come up with experiments which proved conclusively that they were right—under some circumstances, light behaves exactly like a wave, while under others it behaves exactly like a stream of particles. Now we know that both views are right; light is both particle and wave, and so is everything else—although sometimes the wave-particle duality is more obvious, and it can only really be discerned at all for very light objects such as photons (which have no mass at all!) and electrons.

The reason why the wave-particle duality is not obvious in the everyday world is clear from some of the experiments carried out with electrons over the past fifty years or more. When the experiments are arranged to measure the wavelength of the electrons in a "beam," then wavelengths are measured just like wavelengths of radio or light waves. When the experiments are set up to measure the momentum (equivalent to mass × velocity) of the particles in the beam, then momentum is measured, just as if the electrons were a series of tiny cannonballs thumping into the apparatus. And a curious result comes out of these experiments. Whenever the momentum and wavelength are measured, the product (momentum × wavelength) is always the same, a number now known as Planck's constant, after German physicist Max Planck. Planck's constant is very small by everyday standards, 6.626×10^{-27} erg seconds, but the behavior of the electrons is precisely geared to it. If the electrons in the beam are shot into the experimental apparatus faster, so that the momentum of each is doubled, then the wavelength is halved so that the product (momentum × wavelength) is still 6.626×10^{-27} erg seconds. So Planck's constant describes in some very fundamental way the link between the particle nature and the wave nature of things. Because the mass of an electron is just over 9×10^{-28} gram, it is easy to balance the two sides of the Planck equation for electrons and the particle-wave duality is very clear. However, for any everyday object (my typewriter, for example) the mass runs into many grams—even hundreds or thousands of grams—and to balance the equation at all we have to give the equivalent wave packet a ludicrously small wavelength. For all practical purposes, the wave equivalent of my typewriter can be ignored. This doesn't mean there is no wave equivalent for the typewriter, or a car, or a billiard ball, but that the particle nature completely dominates—and all because Planck's constant is so small.*

Fascinating though the implications of quantum mechanics and wave-particle duality are, though, to follow them further would take me a long way from the central theme of this book, and we already have enough information to get back to the way an electron surrounds an atomic nucleus. Treating the electron as a little bundle of wave energy, it is easier to see how it spreads itself around a proton. Like the sound waves which resonate inside an organ pipe, or the vibrations of a plucked guitar string, the elec-

---

* Why Planck's constant should be so small is another matter entirely, and seems to be "just one of those things." If the constant were bigger—say, 1 g sec—then life would be very interesting (if life could exist at all) with typewriters that *did* have to be treated as waves! The links between life, the Universe, and the nature of the fundamental constants of nature will be discussed later in this book.

Hydrogen

1 proton
1 electron
0 neutrons

Helium

2 protons
2 neutrons
2 electrons

Lithium

3 protons
4 neutrons
3 electrons

Carbon

6 protons
6 neutrons
6 electrons

Some of the simplest elements. Although the nucleus, made up of protons and neutrons, lies at the heart of the atom as the Sun lies at the heart of the Solar System, the representation of electrons in orbits like those of the planets around the Sun is not strictly accurate. Because of quantum effects important on the scale of an atom or electron, it is better to think of each electron as "surrounding" the nucleus in three dimensions. But for two-dimensional drawings, the "orbital" representation does at least show how successive electrons are added in "shells" farther out from the nucleus, and indicates how many electrons there are in each shell. Carbon is especially interesting because it "needs" either to add four electrons to fill its outer shell, or lose four electrons to leave a full shell exposed. This is the basis of the enormous variety of chemical reactions on which life depends.

tron "wave" can set up a particularly stable standing vibration, or resonance. Other things being equal, a collection of particles will settle into the most stable state, and an electron wave packet will attach itself to an atomic nucleus in the form of a standing wave surrounding the nucleus. It is impossible to point to one point near the nucleus and say "the electron is here"; all we can do is say that "the energy of the electron is spread over this region of space around the nucleus," and the region of space around the nucleus where an electron sets up its standing wave is called a "shell."

ATOMS, ISOTOPES, AND MOLECULES

So one proton can be surrounded by one electron, in the form of a standing wave of energy, to make a single hydrogen atom. Heavy hydrogen, or deuterium, is just the same except that the nucleus contains a neutron as well as a proton. This is the same element, in chemical terms, because the way an atom interacts with other atoms depends on what the other atom "sees"—and what the other atom sees is one electron surrounding a nucleus. The electron is the accessible part of the atom, and other atoms don't much care how many neutrons there may be buried in the nucleus since all they can interact with are the electrons in the outermost shell of other atoms.

The next element, then, is the one which contains two protons in the nucleus, and therefore has two electrons surrounding it. This is helium, whose nucleus, at least, we have already met. Helium nuclei may contain either one neutron or two, giving two iosotopes (helium-3 and helium-4, named by the total number of particles, protons plus neutrons, in the nucleus). Each nucleus has two attendant electrons, to balance precisely the double positive charge from two protons, but electrons are sufficiently like every-day particles to obey the dictum that two objects cannot be in the same place at the same time, so that they occupy different parts of the shell immediately surrounding the nucleus.

When we move on to the next element, lithium, things get one stage more interesting still. The most common isotope of lithium contains three protons and four neutrons in the nucleus (so is

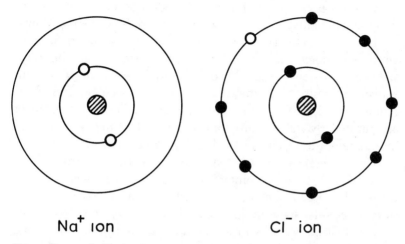

Na⁺ ion   Cl⁻ ion

How sodium and chlorine atoms get together to make sodium chloride, common salt. The single electron in the outer shell of a sodium atom can be slotted into the single gap in the outermost shell of a chlorine atom. This leaves a positive charge on the resulting sodium ion and a negative charge on the chlorine ion. The two stick together because of the electrostatic attraction between positive and negative charges.

called lithium-7), and each nucleus is surrounded by three electrons. But it happens that two electrons occupying a shell right next to the nucleus fill up all the available space; there is no room for a third, and the extra electron has to spread itself around another shell, further out from the central nucleus. Further out, there is more space to play with, and it turns out that there is room for no less than seven more electrons to nestle alongside this third one (making eight in all in the second shell, more or less at the same distance from the nucleus) before the next one has to go into a shell still further out. So the complexity of the chemical elements builds up; for every extra proton on the nucleus, there must be an extra electron which has to fit in somewhere, and tries to get into the most stable shell available. The bigger the atom, the more electrons there can be (generally speaking) in the outer shells, and the chemical properties of elements depend in very large measure on both the overall number of electrons and, in particular, the number in the outermost shell, which are the ones that can interact most easily with other atoms.

With this background of atomic physics it becomes very easy to understand chemistry, and in particular to see how chemical compounds form. The key to making molecules—combinations of atoms joined together—is the number of electrons in the outer shell. For reasons which I won't go into here, but which are explained within the framework of quantum mechanics, configurations in which the outermost shell is exactly filled with electrons are particularly stable, and atoms combine together as molecules in an attempt to reach this particularly stable state. For hydrogen, with just one electron in its only shell, but room for two, this is achieved by two atoms joining forces to make a molecule of hydrogen, H_2. Each nucleus gets a part share in two electrons, instead of a full share in just one electron, and this creates an illusion that both places in the electron shell are filled. Helium, on the other hand, is quite happy with its two electrons and has no predisposition to gain more—so helium is a very inert substance which scarcely reacts chemically with anything.

These simple elements are special cases, by virtue of their very simplicity. The situation as it applies to the molecules of life is better shown by an example from rather heavier elements, and the classic example is sodium chloride, common salt. Each atom of sodium has eleven electrons, which fill the innermost shell (two) and the next shell out (eight electrons), leaving just one electron on its own in the third possible shell. Chlorine, on the other hand, has seventeen electrons, which fill the innermost two shells ($2 + 8 = 10$ electrons), leaving seven electrons for the outer shell, which ideally should contain eight. So sodium and chlorine have a great affinity for one another; the "extra" sodium electron can fit neatly into the "gap" in the chlorine outer shell, and the family of electrons (at least the outermost eight) is shared between the two nuclei, holding them together as a stable molecule of sodium chloride, NaCl ("Na" because the letter S is already used for sulfur, so chemists fall back on abbreviations from Latin or Greek to avoid confusion). One way of thinking about what happens is to say that the sodium atom literally gives up its electron to the chlorine atom; that would leave a net electric charge of $+1$ on the remaining, incomplete sodium atom (called a

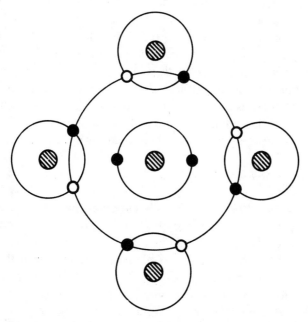

When hydrogen joins up with carbon to make molecules of methane, it is better to think of the outer electrons as being shared rather than swapping over from one atom to another. In this covalent bond, each of four pairs of electrons is shared between the central carbon atom and one of four hydrogen atoms. The result is a molecule of methane.

sodium ion) and an overall net charge of −1 on the corresponding chlorine ion, and the two would stick together by electrostatic attraction. It's a nice simple way of imagining what goes on, and some people find it helps; but I prefer the concept of the outermost electrons being shared by the two atoms, which is a much closer approximation to the idea of electrons as wave packets occupying some spread-out volume of space, rather than little billiard balls slotting into holes.*

---

* Chemists do actually make a distinction between cases where electrons are definitely shared, as with hydrogen molecules, and cases where the "swapover" description makes some sense, as with sodium chloride. The distinction is not important here; what matters is that in seeking to have stable, filled outer shells, atoms try to hold on to "extra" electrons and that by sharing electrons atoms stick together to make molecules.

Calcium is very similar to sodium, but has two "spare" electrons in its outermost shell, and it can share one of these with each of a pair of chlorine atoms, making calcium chloride, $CaCl_2$. And so the complexity of chemical *compounds* builds up. But one element is a very special case, with such a remarkable ability to form compounds with other elements that the study of the chemistry of this one element is a separate scientific discipline in its own right. The element is carbon, and the chemistry of carbon is called organic chemistry, and is intimately linked with life.

CARBON CHEMISTRY

Carbon is special because it has six electrons, two tightly bound to the nucleus in the innermost shell and four in the second shell, where the ideal number would be eight. With a shell precisely half full, carbon has the greatest possible repertoire of chemical activity open to it in the search for stability. It is equally happy "giving up" electrons to other atoms to empty this shell (making compounds such as methane, CH_4) or "gaining" electrons from other atoms with a surplus, to fill up its outermost shell. Or, in my preferred terminology, carbon has four electrons to *share*—with only three in the outer shell, only three would be shared; with five in the outer shell, it would seek three more, again restricting the choice. Each liaison with one other atom is called a bond, and some idea of the arrangement can be seen by drawing out the molecules, with the bonds represented by lines joining the symbols for the different atoms. So methane, CH_4, becomes

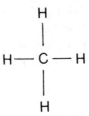

which neatly shows one carbon atom surrounded by four hydrogen atoms, with each of which it shares a pair of electrons. But carbon can also combine with other carbon atoms making, for example, ethane, C_2H_6

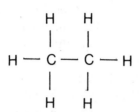

or even ethene, C_2H_4, where four electrons (two "bonds") are shared between two carbon atoms, and four more between each carbon atom and a pair of hydrogen atoms

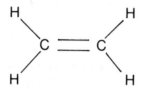

This kind of double bond is not, in fact, stronger than a single bond—if it were, carbon would be happiest as molecules of C_2

$$C \equiv C$$

The shared electrons are too localized in a double bond, rubbing shoulders (or wave packets) with one another, and if hydrogen is available then ethene will tend to grab hold of it and become ethane. The single carbon bond, however, is entirely satisfactory, and whole chains of carbon atoms may hold hands in this way, making complex molecules such as C_4H_{10}

$$
\begin{array}{ccccccc}
& H & & H & & H & & H \\
& | & & | & & | & & | \\
H - & C & - & C & - & C & - & C & - H \\
& | & & | & & | & & | \\
& H & & H & & H & & H
\end{array}
$$

and so on. These chains can double around to make a ring (the most stable ring compounds contain six carbon atoms "holding hands" with other atoms joined on to the outside of the ring) or extend to make a "spine" containing thousands upon thousands of carbon atoms, with a variety of other atoms joined on to the bonds sticking out from the spine. (I have, of course, only stuck to carbon/hydrogen bonds in these examples for simplicity; any of the hydrogen atoms could be replaced by other atoms which are seeking an extra electron, or an atom with capacity to make more bond links may replace more than one hydrogen atom, as in hydrogen cyanide, HCN

$$H - C \equiv N$$

and so on.) The chemistry of life is, essentially, the chemistry of carbon, and in particular the chemistry of very long carbon chains with interesting bits and pieces stuck on to their sides. But—the big question—what exactly do we mean by "life"?

WHAT IS LIFE?

The simplest and best way to distinguish between the living and the nonliving is that living things are able to reproduce—they can make copies of themselves. This definition operates both at the everyday level (a tree is alive; a rock is not) and at the molecular level where life began and where the life processes that control the reproduction of a tree, a human being, or an amoeba operate still today. Life began when—somehow, somewhere—a combination of chemical reactions produced a molecule that was capable of making copies of itself by triggering further chemical reactions. From then on, the story of life—evolution—has been one of competition between different life forms for the available "food" (the chemical elements and compounds necessary to make copies) and protection against other life molecules. So the fundamental molecules of life are today concealed within a protecting wall of material as individual cells, and many millions of cells function together to make up a human being or a tree. How this came about will be the subject of the next chapter—and it is all relatively straightforward once the first life molecules appear. Just how and where life originated, however, is now once again the subject of debate, after a spell of several decades in which no serious challenger to what is still the most widely taught theory had emerged.

Indeed, the conventional theory of the origin of life can be traced back for almost half a century to the suggestion by J. B. S. Haldane that the right organic molecules (carbon compounds) must have built up slowly in the Earth's oceans over a long period of geological time, until the complexity of the compounds being produced by the chemical reactions built up to the point where the first life molecules appeared. (There may even have been a unique "first living molecule" in this picture, from which every living thing on Earth is descended!) It was Haldane who gave the description "primeval soup" to the early oceans, rich in organic compounds, and his name has stuck, although a Russian scientist, A. I. Oparin, suggested a similar theory independently at about the same time, but postulating that life molecules first appeared in hot volcanic pools on land rather than in the broad ocean. One rather obvious—but so obvious that it may be worth stating—point about all this is that there were no life forms around to interfere with the process, by definition! If any complex molecules were created by chemical reactions in the ocean today, they would swiftly be gobbled up as food by some living creature. Life is constantly absorbing simple molecules like those of hydrogen, carbon, and oxygen and combining them into interesting (that is, living) molecules. But when there was no life, molecules could indeed build up slowly without interference, and the first "living" molecule would have a world (or, at least, an ocean) of food at its disposal. How complicated does a life molecule have to be? Well, it *can* be very large indeed, as the example of hemoglobin, the molecule which carries oxygen through our blood, shows.*

Hemoglobin is a complex molecule belonging to the family of substances which are called proteins, and it is made up of chains of smaller molecules, called amino acids, which are twisted together in a shape that is unique to hemoglobin. Each amino

* I have taken this example from Richard Dawkins's book *The Selfish Gene* (London: Oxford University Press, 1976), which I strongly recommend to anyone wanting to delve deeper into the mysteries of the evolution of life.

acid contains a few dozen atoms arranged in a precise pattern on a carbon spine, and in one hemoglobin molecule there are exactly 574 amino acid molecules forming four principal chains which are wrapped around one another to give the hemoglobin molecule its shape, described by Richard Dawkins as "rather like a dense thornbush." A typical healthy human being has 6,000 million million million (6×10^{21}) hemoglobin molecules, all identical to one another, with 400 million million (4×10^{14}) being destroyed every second and replaced by new ones, freshly manufactured by the body and indistinguishable from the rest. Clearly, the human body is no longer operating by the process of random chemical reactions in manufacturing such quantities of identical, complex molecules. Rather, the living body selects chemicals from the store available (from food) and rearranges them in a certain pattern. And, equally clearly, if there were primeval pools of chemicals bubbling away on the primitive Earth, or if the oceans were like a primeval chemical soup, the random chemical processes operating at the time did not throw up molecules as complex as hemoglobin, let alone a living cell. Even if you waited for the entire lifetime of the Earth, 4,500 million years, such an improbable construction as a hemoglobin molecule would not arise by purely random interactions.

This doesn't mean that some guiding hand stuck the chemical bits together in the primeval soup. As soon as molecules that could reproduce themselves—living molecules—appeared, the reactions they were involved in were no longer random, but were using the surroundings to the "advantage" of those molecules, breaking up other compounds and making more replicators. A new natural law immediately came into operation, with the evolution of living molecules being greatly speeded up by the process of natural selection. "Efficient" replicators would soon swamp the primeval soup and turn vast quantities of the available raw materials into replicas of themselves. And any minor changes which did occur by chance—"mutations," if you like—and which made one particular replicator molecule, or a family of replicators, more efficient still at the job of reproducing would allow that molecule, or family of molecules, to literally take over the world. The process must have happened many times indeed to produce human beings and hemoglobin molecules, but the time available is ample for the job once natural selection could operate to rule out inefficient molecular copying and construction. Most people, if they think about evolution at all, regard it as something which happens to whole families of animals and plants, not to individual molecules. But the laws of natural selection which encouraged molecular evolution are exactly the same as those which govern the evolution of species today—and, in a very real sense, evolution today still goes on at the molecular level, rather than at the level of a whole body. The genes which we carry in our cells and which determine the sort of person we are physically (and, to some extent, mentally) are simply very efficient replicator molecules hiding safe inside what Dawkins describes as their "survival machine." Once cells evolved, genes were both protected and forced to manipulate their surroundings by remote control; from that point on, the story of life is understandably the story of "life as we know it." But how did life get to the point where cells were "invented"?*

CHEMISTRY IN SPACE

One thing is now clear from astronomical studies made during the 1960s and 1970s. The "primeval soup," in all probability, started out with a mixture of chemicals already significantly more complicated than the simple water, carbon dioxide, methane, and ammonia mixtures that were the focus of attention immediately after the Haldane-Oparin idea of the origin of life appeared. Many chemists have followed up the Haldane-Oparin idea with experiments in which mixtures of some or all of these chemical compounds are sealed into flasks and treated with ultraviolet radiation

* I do not intend to imply that there was any deliberate choice on the part of replicator molecules to build cells; it happened by chance, aided by natural selection, and because it was a very useful and efficient asset to the job of replicating, cellular life was a great success. However, as any reader who has got this far will have noticed, I have no objection to anthropomorphic descriptions where it helps to make a point and preserve the flow of a concept. It is easier to say "cells were invented" than to say "cell walls appeared as a result of a complex chain of beneficial replicator mutations and, because they conferred an advantage on the replicators cells, soon dominated much of the life on Earth." At the beginning of my version of the story of life, though, it seems appropriate to stress that the anthropomorphism is just that, and that however I may phrase different points in the rest of this book, there is no hint of a consciousness or intelligence involved in the behavior of life on Earth until our immediate hominid ancestors appeared on the scene.

(to simulate the energy of the Sun), or with electric sparks (to simulate the energy of electric storms in the atmosphere of the young Earth), or both. As long ago as 1953, Stanley Miller showed that treating hydrogen, methane, water, and ammonia mixtures in this way for a week produced a yield of 3 percent amino acids (remember the building blocks of hemoglobin? they also appear in many other life compounds) in the mixture. More recently, similar experiments have both produced yet more complicated life molecules and done the trick with initial mixtures more representative of the water/carbon dioxide–dominated atmosphere that was most likely to be produced by outgassing from the primitive Earth. Simple compounds plus energy to encourage chemical reactions produces life molecules, or at least "precursors to life," since no self-replicating molecule has yet been produced in any of these experiments. But nature, it now seems, didn't even need a planet to produce these precursors to life, since simple compounds are scattered right through space in interstellar clouds of gas and dust, where they may receive a great deal of energy from a nearby star. Someone should have twigged that "simple compounds plus energy = precursors to life," but theorists continued to visualize interstellar clouds as cold, dark, and dull places full of only simple compounds—molecular hydrogen, perhaps—and elements—carbon, say—until observations began to show the clouds to be much more interesting places.

Perhaps this is the time to recap briefly on the story so far. After the Big Bang of creation, the Universe contained clouds of hot gas already clumped into a hierarchical clustering pattern as a result of the physical processes which dominated the last stage of the radiation fireball. These clouds contained a great deal of hydrogen, a small proportion of helium, and hardly anything else at all. Most of the matter in these clouds very quickly collapsed into stars, forming supergalaxies ten or twenty times bigger than the galaxies of bright stars as we know them today, and these stars burned nuclear fuel in their interiors, converting hydrogen and helium into more complex nuclei by the process of nuclear fusion. By the time this first generation of stars was flickering out (either with a bang, in supernova explosions, or as a dying whimper) the trace of leftover gas—perhaps only 10 percent of the mass of each original supergalaxy—was settling under the influence of gravity into the middle of the supergalaxy and forming stars in its own right among the swirling gas clouds, which became a spiral galaxy. For some galaxies, collisions and close encounters have distorted the original pattern so that we see a variety of galaxies around us now, with two main families, the spirals and ellipticals. Our own Milky Way Galaxy, however, seems to have avoided major interactions of this kind, and remains as a spiral embedded within a now dark array of ten times as many dark stars or black holes.

Within this spiral galaxy, the processes of star formation, nuclear synthesis, and stellar explosions have continued. So between the stars there are clouds rich in the elements produced by nucleosynthesis inside stars, including carbon, nitrogen, and oxygen, as well as a great deal of unprocessed original hydrogen (a great deal by human standards, an almost insignificant remnant compared with the original glory of our supergalaxy). So it was no surprise to astronomers to find, from analysis of the spectra of light from the vicinity of interstellar clouds, ample evidence for the presence of a variety of atomic elements, and a few simple compounds such as CN and OH. (OH is strictly the hydroxyl "radical," not a proper molecule; oxygen would "like" to have another hydrogen atom, making H_2O, water, but it will make do with one if that is all it can get.) Since anything in an interstellar cloud can be part of the input to a new solar system forming out of such a cloud—our own Solar System formed in just this way—it seemed reasonable to put such simple compounds and elements in as the starting point for the evolution of life on Earth.*

This is where things stood in 1968. Astronomers knew that many elements and a few simple compounds existed in interstellar clouds, biochemists accepted these as inputs to the processes which produced life on Earth, and experiments had shown that precursors to life could be made by putting energy into closed sys-

---

* I am fond of the strictly accurate astronomical references in the song "Woodstock" to the fact that we are made of "stardust, billion-year-old carbon," since it is literally true that everything in our bodies except hydrogen has been processed through at least one star, and at least a billion (thousand million) years ago. The hydrogen has had a much duller existence—although some may have been involved in nuclear reactions inside stars and been spat out again as protons, some of the hydrogen on Earth, and therefore in our bodies, has been unchanged since the Big Bang itself.

tems containing simple compounds. Then Charles Townes, a Nobel Prize–winning physicist, came onto the astronomical scene. Townes was interested in radioastronomy at short wavelengths, the few centimeters or so wavelengths which are called microwaves and are widely used in communications links, radar, and so on. These wavelengths are just the ones where many fairly complex molecules produce or absorb electromagnetic radiation. This process is the electromagnetic equivalent of the resonance of a tuned guitar string, and just as a shorter string produces a higher-pitched note (that is, one with a shorter sound wavelength), so a smaller molecule resonates with a shorter wavelength of electromagnetic radiation. Individual atoms, indeed, are so small that they "resonate" with light waves, and this is what produces the lines which distinguish their spectra: a dark line at a precise wavelength shows light being absorbed by the atoms of one element, a bright line somewhere else shows light being radiated by atoms of another element. Large molecules produce the equivalent of "lines" in the microwave "spectrum"—either peaks of radio noise or wavelengths where relatively little noise is heard because radio energy from behind the molecules is being absorbed before it can get to the receiver.

There are two snags, however, in looking for complex molecules—"polyatomic" molecules—by tuning radio telescopes to investigate microwave radiation from space. First, the shorter the wavelength the more accurate the antenna and receiver system has to be. Putting it very simply, if you are looking for radiation with a wavelength of a centimeter or so, then the surface of the antenna has to be smooth to within a centimeter, or the incoming radiation gets bounced off in any direction from what is, to it, a rough surface. Optical telescopes have to be smooth compared with the wavelength of light, substantially less than a millionth of a meter. It is difficult to make a large, smooth surface, yet for weak radio noise coming in from space we need the biggest possible antenna to catch as much radiation as possible. So a microwave antenna is a compromise between size and consistent smoothness. The receiver's amplifier system then has to boost the weak signal to detectable levels without destroying the information it contains, and it was only in the 1960s that suitable receiver/antenna systems were developed for this work, building in large measure on the expertise gained from the first broadcasts using communications satellites. (Penzias and Wilson, another pair of Nobel Prize winners, made their pioneering discovery of the cosmic background of microwave radiation using a purpose-built satellite communications system.)

And the second snag, still a problem today but a much bigger one in 1968, is that although the microwave radio "lines" can identify a polyatomic molecule as precisely as a fingerprint identifies a person, it is first necessary to "take the fingerprint" by making the appropriate compound in the laboratory and studying its microwave spectrum firsthand. The more atoms a molecule contains, the more subtly different electromagnetic resonances it is likely to be involved with, making the fingerprint distinctive but the effort needed to measure it tedious. It wasn't until some complex molecules had been discovered in space that the effort of identifying others seemed worthwhile—nobody wants to spend hours in the laboratory determining the microwave fingerprint of a molecule, then go and make the radio observations only to find that the particular molecule isn't there! So when a Berkeley team inspired by Townes set off on the trail of polyatomic molecules in interstellar space in 1968, they started small. Their dramatic success soon encouraged others, however, and the whole business took off in a big way as more and more molecules were identified in space.

First came ammonia (NH_3), identified in December 1968 by its radiation at a wavelength of 1.26 cm. Hardly really a polyatomic molecule, but it showed the system worked, and the molecule had never been identified in space before. Then the same team found the microwave fingerprint of water in space, and early in 1969 the cat was put among the astronomical pigeons when another team found a pattern of radio "lines" which exactly fitted that of the molecule formaldehyde, one of the few to have been studied in this way on Earth. Formaldehyde is not just a complex molecule by the standards of molecules known in interstellar clouds at that time; it is an organic molecule (H_2CO), one of the enormous family of carbon compounds and, specifically, one of the molecules which can easily be made by adding energy to a mixture of simple compounds. At this point the penny dropped, at least as far as some astronomers were concerned. (I was going to say "astrophysicists," but I suppose 1969 marked the beginning of a new subbranch of astronomy, "astrochemistry.") Simple compounds in interstellar clouds, plus energy, could surely make the precursors of life as efficiently as on Earth, with the bonus that whereas the history of the Earth covers no more than 4,500 million years, a cloud in space might have been around for thousands of millions of years before that, doing the mixing job required before life itself came on the scene and greatly shortening the interval required for life to develop on Earth. If the Earth was "seeded" by chemicals of this complexity early in its history (perhaps from the impact of comets, which are thought to contain almost unsullied "raw material" of the interstellar medium), then the precursors of life would be present even in the earliest primeval soup, and it wouldn't require long, compared with the age of our planet, for replicators to appear.

LIFE FROM SPACE?

Dozens of polyatomic molecules have now been identified in space,* and almost all of them are carbon compounds, entirely in line with the known chemistry of carbon and its prolific ability to combine in different ways with many other elements and with itself. Some of these molecules, including formaldehyde, have now been detected even in the radiation reaching us from other galaxies, and there is every reason to believe that carbon-based molecular complexity is a common feature, not just of the Earth or even of our Galaxy, but of the Universe at large.†

* A paper in *Nature* (February 1980) by Mann and Williams listed ninety interstellar molecules identified at that time.

† I cannot resist giving myself a pat on the back by quoting from one of the first books I wrote, *Our Changing Universe*, published in 1976 and written just as some of these discoveries were being made: "We may never know for sure that there is other intelligent life in the Universe, but the odds now seem to favor it. Far from our own planet being unique in this respect, the discovery of molecules like formaldehyde in other galaxies suggests that even our own Galaxy may not be unusual in possessing planets on which intelligent beings live. It is a curious and, to my mind, rather comforting thought that somewhere in the galaxy NGC253 there may be carbon-based life forms with DNA in their cells, studying the heavens with their radio telescopes, and that they may be speculating on the significance of the detection of formaldehyde emission from the rather unspectacular collection of stars which it pleases us to call 'our' galaxy."

So at the very least the young Earth must have been laced with molecules such as H_2CNH, $HCCCN$, H_3CCOH, and others; furthermore, some of the molecules already identified are known to combine easily with one another to make yet more complicated molecules, and formic acid ($HCOOH$) and methanimine (H_2CHN) in particular, both identified in dense clouds in space, react to produce the simplest amino acid, glycine (NH_2CH_2COOH). Formaldehyde, the key discovery in all this excitement, is itself a common component of bigger organic molecules, including sugars, which are essential to life as we know it. Quite probably, even amino acids were present in the primeval soup, a revelation which prompted Jim Lovelock to say that "it seems almost as if our galaxy were a giant warehouse containing the spare parts needed for life,"* and to draw an intriguing analogy with a planet made up entirely of the components of watches. Given a long enough time—perhaps a thousand million years— argues Lovelock, tidal forces and the movement of the wind will assemble at least one working watch. Given a planet rich in the components of life, chemical reactions will produce a replicator molecule within a thousand million years or so. The odds against the sequence of chemical reactions that produced the first replicator are astronomical, but so was the time available. As Lovelock puts it, "Life was thus an almost utterly improbable event with almost infinite opportunities of happening," so it happened eventually. For the rest of my discussion of where we came from, this, the "establishment" view today, is adequate. What matters, as far as you and I are concerned, is that at least one replicator did form and that evolution has proceeded since that time to produce the variety of life around us today. But the established view is still not entirely satisfactory, and it is natural to wonder about just what went on for thousands of millions of years in those interstellar clouds laced with the precursors of life and supplied with plenty of energy from nearby stars. Happily for those of us with fertile imaginations who wondered in this way after the discovery of the complexity of interstellar molecules, Fred Hoyle and Chandra Wickramasinghe have done more than just wonder about the

---

* Quote from Lovelock's book *Gaia: A New Look at Life on Earth*; more of his ideas later.

problem, and have come up with a detailed theory of how genuinely living forms—replicators—may have become established in interstellar clouds.

The implications make the mind boggle, but in a pleasant sort of way. If the kind of replicators we are descended from started replicating in interstellar clouds, then it is all the more likely that life everywhere is descended from the same sort of replicators. If our planet was seeded with replicators almost at birth, then even the problem of making a watch out of its constituent parts just by waiting long enough is removed. Anyone can *speculate* about ideas of this kind, but Hoyle and Wickramasinghe have gone much further than this, offering an explanation not just of how but also of why conditions in interstellar clouds may favor the development of replicators—of life. Although the theory is still new and will surely be modified, perhaps drastically, as new discoveries are made, I am persuaded of its validity as a broad description of the origin of life and, admitting that I was predisposed in its favor (as the quote above shows!), offer it here as the best available answer to the question of where the first replicator came from.

Most of the material between the stars is simply hydrogen gas, truly primeval material left over from the epoch just after the radiation-dominated era of the cosmic fireball. But the most obvious material between the stars, showing up as dark patches on astronomical photographs, is dust. These dust patches literally block out the light from stars behind them, so that often the densest patches appear like black pits against the surrounding sky dominated by bright stars; but even where there is no patch of dust in the line of sight dense enough to block out the light entirely, hardly any starlight reaches us without being affected by more tenuous regions of dust which it encounters on its way here. As well as being dimmed a little by the ubiquitous dust in space, starlight is reddened slightly, in exactly the same way that dust in the atmosphere of the Earth causes red sunsets and sunrises. Shorter wavelengths of light—the blue end of the spectrum—are more easily scattered by small dust particles, while longer wavelengths—the red end of the spectrum—can get through a collection of small particles without being so greatly affected. So light looks redder after it passes through a cloud of dust in space, or the hazy dust of the lower layers of the atmosphere, in the case of spectacular sunsets.*

About 2 percent of all the mass of interstellar clouds is dust, with the rest mainly hydrogen gas, plus helium, forming great clouds which are the birthplaces of new generations of stars. The dust in interstellar material is a product of the nucleosynthesis in earlier generations of stars, which we know produces carbon, oxygen, and nitrogen, so the dust grains in interstellar clouds must be made of combinations of these elements with one another, or with hydrogen. (Remember that helium is very stable, "needs" no extra electrons and has no surplus to give up, and scarcely reacts with anything.) The amount of dimming in starlight at particular wavelengths in the spectrum shows just how big the particles that are doing the dimming are, and since the 1930s it has been known that these dust particles must be tiny grains about the size of the wavelength of visible light. This is measured in Ångstroms (Å), and one centimeter is 100 million Ångstroms, while optical light covers a band of wavelengths around a few thousand Å; so the size of an individual interstellar dust grain is less than one hundred-thousandth of a centimeter.

Even so, it is important to appreciate just how much of this interstellar material there is in our Galaxy. Including all the gas, interstellar matter makes up about one-tenth of the mass of the Galaxy, which amounts to 10 thousand million (10^{10}) times the mass of our Sun.† Two percent of this is still an impressive 200 million solar masses of material scattered around our Galaxy in tiny dust particles in interstellar clouds.

Since the 1930s, observations of the amount of dimming produced at different wavelengths—the amount and nature of the red-

---

* This reddening is quite different from, and has nothing to do with, the Doppler "red shift" which bodily changes the wavelength of light from a galaxy receding from us at high speed. Indeed, because some distant objects in the Universe, the quasars, produce a lot of energy in the ultraviolet part of the spectrum which is normally invisible, and the red shift pushes this energy peak into the blue part of the optical spectrum, the light from a quasar with a high *red shift* is often very *blue*, and this is one way astronomers search for previously unknown quasars—by examining photographic plates for very blue starlike objects. Dust-reddened light is *red* because the blue component has been scattered out and the remaining spectrum is dominated by the red.

† I refer here, of course, to the bright stars of the Milky Way Galaxy familiar from optical photographs, which is itself only one-tenth of the mass of the supergalaxy in which, according to the best interpretation of the latest evidence, the bright galaxy must be embedded.

dening—have been extended and improved, and today we even have measurements of the absorption by interstellar radiation beyond the optical spectrum, at infrared and ultraviolet wavelengths. It turns out that the strongest absorption (measured by instruments carried on rockets and satellites above the Earth's atmosphere, which itself blocks out ultraviolet light) is at around 2,200 Å, in the ultraviolet band of the spectrum. Before this discovery was made in the mid-1960s, the most widely favored explanation of the nature of interstellar grains saw them as icy particles, a kind of "snow" of frozen water, methane, and ammonia. Even before the ultraviolet measurements came in, however, Hoyle and Wickramasinghe were arguing that it is very difficult to make ice grains or snow in space, and that it would be a lot easier to get dust clouds made up of tiny particles of carbon—soot.

This may seem bizarre. But carbon is one of the major products of stellar nucleosynthesis, and there is a family of stars called carbon stars, which are shown by their spectra to have atmospheres rich in carbon, and which vary regularly with periodic fluctuations roughly a year in length.* All stars are constantly losing matter from their atmospheres, and variable stars can be thought of as "puffing" this stellar wind outward with the same periodic cycle as that of their brightness variations. So Hoyle and Wickramasinghe came up with the graphic picture of a family of sooty stars coughing out carbon into space—an idea backed up by later observations which revealed spectral features corresponding to graphite (carbon particles, like the inside of an ordinary pencil) in the radiation from carbon stars.

The discovery of the strong absorption at 2,200 Å really rules out the possibility of icy interstellar grains, since none of the three candidates (water-ice, ammonia-ice, or methane-ice) blocks radiation particularly strongly at this wavelength.† On the other hand, carbon does—so well, indeed, that some astronomers argue that *only* carbon grains are needed to explain the interstellar reddening.

* If we see a category of stars which we can call "carbon stars" today, the natural explanation is not that these are unusual stars, but that all stars go through a phase of being carbon stars during their lifetimes, with only a few visible in this state at any one time.
† Later observations have shown an absence of absorption where the ices *should* affect the light coming through to us, and the infrared radiation also indicates temperatures in many interstellar clouds above the boiling point of water. The case seems proven!

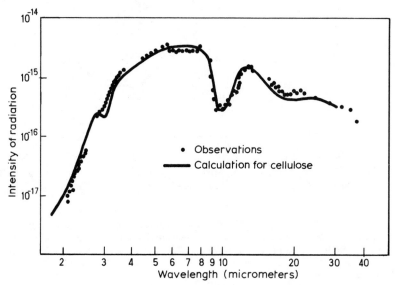

To consternation and, it must be said, disbelief among the astronomical community, Fred Hoyle and Chandra Wickramasinghe have presented evidence that even so complex an organic molecule as cellulose may exist in clouds in space. The observations of absorption by interstellar clouds at infrared wavelengths exactly fit the theoretical curve calculated for cellulose. Is this a coincidence?

But there must be nitrogen and oxygen present in the clouds, too, quite apart from the hydrogen gas in which the dust is embedded. There "ought" to be molecules present, and Hoyle and Wickramasinghe have shown just which molecules can best explain the observations, and how they may have formed.

At this point, the story switches to studies of the infrared radiation from clouds of dust and gas which are actually collapsing to form stars, or from the "cocoons" of dusty material surrounding young stars. These clouds are heated by the energy coming out from the forming, or newly formed, stars inside, and reach temperatures of several hundred degrees, which produces strong infrared—heat—radiation from the clouds themselves. The peak intensity of the radiation comes at a frequency in the infrared which corresponds both to the actual temperature of the cloud and to the nature of the particles in the cloud—in physics jargon, the underlying black body curve (remember the 3 K background? This radiation is also black body, but at temperatures of hundreds

The relatively simple ring-shaped molecule formed by five carbon atoms and one oxygen atom, the pyran ring, tends to make replicas of itself, linked by extra oxygen atoms to form the skeleton of a polysaccharide chain, if more carbon and oxygen atoms are available. This form of complex, not-quite-living molecule could easily be built up under the conditions which we know exist in interstellar clouds.

K) has superimposed on it the spectral emission features of the particles.

When we move into the infrared and wavelengths increase, the traditional Ångstrom unit becomes rather unwieldy as a measure of wavelength, and infrared astronomers prefer to use the micrometer (one micrometer is a millionth of a meter, 10^{-6}m), which is, in any case, a rather more logical unit directly tied to the metric system of measurement. Whereas a micrometer is a millionth of a meter, an Ångstrom is one ten-thousand-millionth of a meter (10^{-10}m), so that there are 10,000 Å in one micrometer. Features in the infrared band may be found at wavelengths of a few tens of micrometers, which in Å would give us numbers above 100,000 to play with, so I'll stick with micrometers! (But if you want to, convert to Å simply by adding four more zeros to all the numbers given as micrometers.)

Three strong infrared features dominated the spectrum from many of the hot clouds of dusty material associated with the birth of stars. These come at two to four micrometers, eight to twelve micrometers, and around eighteen micrometers. Theories have been devised to explain these by three different families of chemical compounds in the clouds, including variations on the old water-ice idea and the presence of silicates. But seeking the simplest possibility, Hoyle and Wickramasinghe tried to find one substance to account for the whole pattern. To their own amazement, as they put it, and to the consternation of many other astronomers, they found the ideal candidate—cellulose, which just happens to be the most abundant organic material on Earth, a fundamental part of the structure of plants!

LIFE IN SPACE?

It is only fair, at this point, to say that a great many astronomers (and biochemists) find the possibility of interstellar cellulose completely unacceptable, and that other theories (rather contrived, but workable) do still exist to explain the infrared observations. The Hoyle-Wickramasinghe hypothesis is not yet the established view, but it provides the most complete explanation of what is going on in the dust clouds of space, and I believe that even if the details are changed by later discoveries, something along these lines will eventually become the established view. Cellulose itself is a member of the family of large molecules called polysaccharides, and in a more recent refinement of the Hoyle-Wickramasinghe idea they have now shown that all the infrared observations of this kind can be explained by a combination of polysaccharides (always the *same* combination of polysaccharides) plus some hydrocarbons (carbon-hydrogen compounds, just what you'd expect in a hydrogen cloud containing carbon grains). So how could a very large molecule with a complex structure arise under the conditions of interstellar space?

As I have already hinted, it turns out that conditions well suited for the development of prebiologic chemistry exist in the outflow of hot dusty material being blown away from young stars as they settle down into their nuclear-burning lives. This dust, of course, comes from a *previous* generation of stellar activity, and is the leftover material from the cloud out of which a cluster of new stars has formed. The energy of the radiation from the star or stars inside the cocoon keeps the carbon and oxygen in the middle of the cloud mainly in the form of single atoms, since strong radiation can break up any molecules (say, of carbon monoxide, CO) that try to form. But where the temperature drops below about 1200 K carbon and oxygen can get together—not as carbon monoxide, but in chains built up from the pyran ring, each ring made of five carbon atoms and one oxygen atom "holding hands" in a circle (C_5O). One pyran ring links easily to another through an extra oxygen atom which bonds to one of the carbons from each ring, and a chain of pyran rings linked in this way forms the skeleton structure of a polysaccharide. And although it is a tedious process to produce one such chain by sticking atoms together, once formed, a pyran ring shows one of the fundamental qualities of life—it acts as a pattern, or template, which encourages the formation of more rings, which link up in a growing chain. When the chain splits into two or more pieces, each piece then continues growing, pulling carbon and oxygen out of the surrounding chemical mixture to make more pyran rings. Now although nobody would argue that a simple pyran ring is alive (its behavior is more like that of a growing crystal—given precisely the right mixture of chemical elements, it will stick them together as an extension of itself, but it does not manipulate its environment by breaking up other molecular compounds and rearranging them into distinct, self-replicating units), this behavior ensures that under the right conditions a great deal of the carbon and oxygen available is turned into polysaccharides, just as under the right conditions a great deal of carbon, nitrogen, and oxygen on Earth has been turned into living cells.

Where is the nitrogen in interstellar clouds, which ought to be there if our understanding of stellar nucleosynthesis is correct? Getting back to the reddened light which comes to us through cool clouds of dust in space—cool clouds which, presumably, include material blown away from young stars, in which the polysaccharide-building process has taken place—there are still many spectral features to be explained. One such, the most prominent, is in a band around 4430 Å and stretches over a width of 30 Å. The width of the feature indicates in a crude way the size of the molecules producing it—atoms make sharp lines, large molecules produce broad spectral features— and this particular feature can be explained by the presence of a particular large molecule, $MgC_{46}H_{30}N_6$, or a family of similar molecules, the porphyrins.

And where else do we find porphyrins? They are some of the fundamental building blocks of the even more complicated chlorophyll—a fundamental chemical involved in photosynthesis, the process by which plants turn sunlight energy into food. Since all animal life depends on eating plants to survive, chlorophyll is arguably the most important biochemical of all in human terms! Hardly anyone takes the "explanation" of the 4430 Å feature in terms of porphyrins seriously; but there is no better explanation on offer, and it certainly provides food for thought, if nothing else. As Hoyle and Wickramasinghe have commented in their book *Lifecloud*, "The union of four C_4N rings to form the central core of a porphyrin molecule is energetically profitable, and for this reason we may expect a fraction of material in nitrogen-rich mass flows [out from young stars] to condense into such molecules." [*]

The wealth of prebiologic material which could be present in interstellar clouds is astonishing. But how could any of it survive the turmoil of the formation of the Solar System and the Earth? The answer is it couldn't—but it might be brought into the inner Solar System very soon after the Earth formed, by visitors from deep space, comets. Comets occasionally produce spectacular displays in the sky, when one of their number dives in past the Sun and produces a stream of material, a glowing cloud which can stretch far across the sky. The most famous, Halley's comet, is due back in February 1986, although it may lack the glory of previous visits as material has slowly been boiled away from it each time it passes near the Sun. Halley's is a regular visitor with a period of seventy-six years; others, however, visit us only once as they swing in from the depths of space, and these are the ones which may have carried the seeds of life to Earth.

As much as 100,000 million comets' worth of material may surround the Solar System at a distance of about one light-year, with pieces sometimes disturbed, perhaps by the gravity of other stars, to plunge down through the Solar System, whip around the Sun, and head off into space again. Each comet may have no more than one thousand-millionth of the mass of the Earth, small by astronomical standards but enough to produce a

* Fred Hoyle and N. C. Wickramasinghe, *Lifecloud* (London: Dent, 1978), p. 98.

noticeable impact should it hit the Earth. The famous "Tunguska Event," a major explosion in Siberia in 1908, was almost certainly a cometary impact, which fortunately hit an unpopulated region of our planet. A similar strike on a major city could kill millions, and the extreme implications of a cometary impact have been described in fictional form by Larry Niven and Jerry Pournelle in their excellent book, *Lucifer's Hammer*. The immediate origin of the cometary cloud is not known—some astronomers argue that it is leftover material from the formation of the Solar System, others that it is not always present but is made up of a temporary influx of material picked up when the Solar System passes through an interstellar cloud of dust and gas. But it is certain that it contains a great deal of material more or less in the state it is in in those interstellar clouds where such interesting molecules seem to be present.

Another trace of material reaches the Earth from space in the form of meteorites, rocky fragments, some rich in iron, which penetrate to the ground. Many more particles, too small to survive the passage through the atmosphere, burn up as meteors ("shooting stars"), while the finest particles of all can float down through the air to the ground. We are continually receiving visitors from space, and any chemical compounds "out there" must reach the Earth from time to time. And this includes another surprise—the discovery in some meteorite samples of amino acids, the fundamental constituents of protein (remember hemoglobin?). The average content of amino acids in meteorites is only about fifteen parts in a million—but this is a much higher proportion than even the most optimistic proponent of the Haldane-Oparin version of the origin of life could expect to have formed through the effect of ultraviolet light and thunderstorms on a primeval ocean soup.

COMETARY PONDS

Throwing caution to the winds, Hoyle and Wickramasinghe go to the opposite extreme. They argue, first, that most of the atmosphere and oceans of the Earth result from the impact of comets when the Solar System was young and the comet cloud thicker (this implicitly assumes comets are left over from those days; it seems to me more likely that the comet cloud is renewed every time we cross the dust lanes of a spiral arm, since surely any original comet cloud would be disrupted by such encounters); then they suggest that genuine life forms may have evolved in cometary clouds even before the precursors of life were brought to Earth by further collisions with comets. The "primeval soup," they argue, might well have been inside a comet, where polysaccharides, porphyrins, and the rest could build up into living, replicating forms.

Another link in the chain joining molecules identified in space with the origins of life on Earth was closed early in 1980, when W. M. Irvine, S. B. Leschine, and F. P. Schloerb published in the scientific journal *Nature* the results of some new calculations about the conditions likely to exist in the cometary cloud which surrounds our Solar System. The greatest single objection to the Hoyle-Wickramasinghe idea, they pointed out, had been that out there in the depths of space comets ought to be frozen, whereas life molecules could build up most easily in liquid water which dissolved organic compounds to make a primeval soup (or, as Charles Darwin himself put it, a "warm little pond").* Hoyle and Wickramasinghe have disputed this, arguing that carbon grains might encourage molecules to stick to their surfaces, building up complex compounds in this way; but Irvine's group from the University of Massachusetts has now removed the original problem anyway.

The solution is almost ludicrously simple. Although conditions in the cometary cloud are cold enough to freeze water unless some source of heat is available within the cometary nuclei themselves, as we saw earlier in this book it is now virtually certain that the formation of the Solar System was triggered by a nearby supernova explosion, which as one of its products left a scattering of radioactive elements in the presolar nebula. If the comets are leftover ma-

---

* Darwin's actual words are quoted by Clair Edwin Folsome in his book *The Origin of Life* (San Francisco: Freeman, 1979): "It is often said that all the conditions for the first production of a living organism are now present, which could ever have been present. But if (and, oh, what a big if) we could conceive in some warm little pond, with all sorts of ammonia and phosphoric salts, light, heat, electricity, etc., present, that a protein compound was chemically formed ready to undergo still more complex changes, at the present day such matter would be instantly devoured, or absorbed, which would not have been the case before living creatures were formed." So Darwin was way ahead of even Haldane and Oparin; but even he never envisaged the warm little pond as part of the nucleus of a comet!

Comets are thought to consist of icy agglomerates of material, frozen by the cold of space. The sometimes spectacular tails that they "grow" when they pass near the Sun on a passage through the inner Solar System are probably the result of material vaporizing and streaming away from the "head" of the comet. But were comets always purely ice? One recent suggestion is that early in the life of the Solar System comets may have been warmed by the radioactive decay of unstable elements left over from the supernova explosion that triggered the formation of the Solar System. In that case, with molten water at their hearts, they could have provided the "warm little ponds" in which life first arose, and from which it was spread to Earth by comets passing nearby and streaming material across space, or from the direct impact of a comet with the young Earth. (*Photograph from the Lick Observatory*)

terial from the original collapse of the Solar System, they would contain radioactive elements from the supernova, and in particular the isotope aluminum-26. The radioactive "half-life" of aluminum-26 is 700,000 years, and as it decayed early in the history of the Solar System the heat produced would have warmed the hearts of cometary nuclei, melting water to provide exactly the warm little pond that the evolution of the first living molecules may have required—or, rather, 100 thousand million separate warm little ponds, with 100 thousand million chances for life to develop.

If so, this is where the cell was invented, the "house" which protects the replicator molecules from their surroundings while allowing them to take in "food." But even if you do not accept this extreme view of the Hoyle-Wickramasinghe hypothesis, it would have been a lot easier, and a lot quicker, to build up living cells on Earth starting out from prebiotic molecules carried in by comets than from simple carbon dioxide, water, and ammonia. Evidence from the rocks of the Earth's crust shows that organisms like bacteria and blue-green algae were present on Earth as much as 3,100 million years ago, which is less than a thousand million years younger than the oldest known rocks, and only about 1,500 million years after the Earth cooled and solidified. It took a very long time for these simple life forms to evolve into the complexity we see about us now; is it reasonable to expect all of evolution up to that stage to have occurred on Earth in only a thousand million years, or more reasonable to place that crucial chemical activity in interstellar clouds where it could build up over many thousands of millions of years before the Solar System formed? Whichever theory gets your vote, from here on my description of the origin of life is very much in the mainstream of current scientific thinking, from the appearance of single-celled life forms to the emergence of man.

6 THE ORIGIN OF SPECIES

It is important, perhaps, to stress that all of the ideas concerning the origins of the first replicators are, to some extent, speculative. We do not *know* whether the first replicators appeared in a "warm little pond" at the heart of a young comet or in the shallow seas of the young Earth. Indeed, although all life on Earth today is built upon the same basic replicator mechanism, involving the life molecule DNA, we do not know whether this was the original replicator system or whether some other system developed first and was superceded by the DNA system at a relatively early stage. What matters, though, is that the modern understanding of the evolution of the Universe, stars, and planets is not at all embarrassed by the existence of life, but can explain in a general way how replicators came into existence, wherever their warm little pond was located. It is no surprise to modern astronomy to find life like us on a planet like the Earth. And while the lack of direct evidence may make it impossible ever to determine exactly how life first got a grip on Earth, we now have a very clear idea of how life developed—evolved—from the state for which we do have the earliest direct evidence up to the present day. This covers more than 3,000 million years of the Earth's history (perhaps 20 percent or less of the history of the Universe so far) and the development of life from single-celled organisms to the variety of the present day, including human animals with many billions of living cells co-operating to produce one distinct living organism, endowed with mobility, a perception of the surrounding world, and, not least, a self-awareness which leads directly to the question "where do we come from?" All this results from the natural processes of selection operating among replicators—some replicators are more efficient at reproducing than others and some survive better than others. Over 3,000 million years, natural selection has led to the diversification of species and to the production of multicellular organisms; but "old" biological systems are not replaced where they continue to reproduce effectively in their own ecological niches, and alongside the "modern" multicellular organisms we can still find types of single-celled species descended almost unchanged from the first colonists of the Earth 3 thousand million years ago.

This indicates the dichotomy of the "struggle for survival": nothing *wants* to evolve, and the basic life process is that of replication, as accurately as possible, of existing molecules. The success of replication at this level is shown by the "living fossil" single-celled species that remain unchanged since the beginning of the story of life on Earth. Changes only happen by mistake, and very few of the copying mistakes are beneficial. Most imperfect copies of replicators do not survive, but end up as chemical "food" for successful replicators. Very occasionally, however, an imperfect copy turns out to be better at the job of converting chemical food into replicas of itself than the original—and such rare mutations not only survive but spread through the environment. Over many millions of years, the accumulation of such rare beneficial copying errors gives rise to species as diverse as a mouse and a mushroom. But the process happens willy-nilly, and in the replicating game the single-celled varieties that have been unchanged for thousands of millions of years could, from one point of view, be regarded as more "successful" than the collection of bizarre mistakes that has produced you and me.

THE GENETIC CODE

To understand just how the replication process produces the mistakes which lead to the origin of species, we need to know a little about the life molecule itself, DNA. Deoxyribonucleic acid (DNA) is the basic copying material for essentially all life on Earth. A bacterium, a mushroom, a blade of grass, or a man are all built according to specifications laid down in molecules of DNA within their cells, and reproduction takes place by separating special cells off, with copies of the relevant DNA molecules, to build up new organisms (at least, this happens for multicellular organisms; the single-celled varieties just split into two cells, each carrying a set of DNA molecules). The DNA molecules which carry the "plans" for a whole organism (such as a human body) are called chromosomes; specific bits of the chromosomal DNA, which might carry such detailed information as whether the body should have blue eyes or brown, fair skin or dark, are called genes, and we shall hear a lot more about them later. Obviously the chromosomal DNA of a human being is considerably modified from the original DNA of the single-celled replicators from which we are descended—otherwise we too would be single-celled replicators. But since I expect most of the readers of this book to be human beings, it is most relevant to us to see the complexity of DNA language by looking at the case of human replicators.

Every DNA molecule is built in a double helix structure, with chemical bonds linking molecules up each spiral, and pairs of molecules linking across from one helix to the other, producing a structure very much like a spiral staircase, with the cross bonds making up the rungs.* Each building block of DNA—the chemical subunits which combine to build up the spiral staircase—is called a nucleotide, and there are only four basic nucleotides combined into every living DNA molecule. The names of these nu-

---

* A minority of present-day molecular biologists argue that DNA is not made up of two long molecules twisted around one another to make a double helix, but of two long molecules each twisted into a long single helix, lying side by side and joined up like the opposite sides of a closed zipper. If this idea is correct, the strands must be built up from both left-handed and right-handed spirals, whereas the double helix picture implies that all DNA molecules are left-handed. This minority view, championed by Dr. G. A. Rodley of the University of

cleotide building blocks are abbreviated as A, T, C, and G; and as the four blocks occur in different orders along the length of a gene or chromosome, it is as if the plans for building and maintaining the whole living organism are written out in a four-letter alphabet. This, if you like, is the "language" of DNA; and all living organisms on Earth share the same four-letter alphabet and DNA language, convincing evidence that we are all descended from one uniquely successful ancestor, whether that ancestor first appeared inside a comet or in Haldane's terrestrial primeval soup. A four-letter alphabet might seem restrictive, but today modern computers are based upon the much simpler language of binary arithmetic, a two-letter language in which the only switching possibilities are "on" and "off," the only answers for any question "yes" or "no." We all know how successful modern computers are, and it is by comparison with their binary language that the power of the DNA four-letter alphabet becomes apparent.

If the genetic code of chromosomes were written in binary language, the number of "bits" of information in a chromosome would be simply twice the number of nucleotide pairs matched across the molecule's "spiral staircase." With a four-letter alphabet, the number of bits of information (yes/no answers) that can be packed in is four times the number of nucleotide pairs. And a single chromosome may contain 5,000 million nucleotide pairs, while each human cell contains forty-six chromosomes. How much information is contained within one chromosome's 20,000 million $(4 \times 5,000$ million) bits? Carl Sagan, in his compelling book *The Dragons of Eden*, made a neat analogy along the following lines:

Consider the human language expressed in the modern version of the Western alphabet with its twenty-six characters and ten numbers. To specify any one letter of the alphabet in terms of the yes/no binary code, we need to ask a series of questions, which Sagan spells out in the case of the letter J:

Canterbury in New Zealand, seems to fit the observations, which depend on X-ray studies of DNA, at least as well as the double helix version. But the X-ray diffraction techniques are not good enough to settle the issue one way or the other. For the record, though, it does seem that it would be easier to do the "unzipping" necessary for the DNA molecules to be copied if they are paired side by side—the double helix has got to unwind, as well as unzip, to do the same job. And copying, after all, is at the heart of the life process.

1. Is the character a letter (answer 0) or a number (answer 1)?
   Answer: 0
2. Is it in the first half (0) or second half (1) of the alphabet?
   Answer: 0
3. Out of thirteen letters in the first half of the alphabet, is it in the first seven (0) or the last six (1)?
   Answer: 1
4. Out of the letters H, I, J, K, L, M, is it in the first three (0) or the last three (1)?
   Answer: 0
5. Out of the letters H, I, J, is it H (0) or is it one of I, J (1)?
   Answer: 1
6. Out of I, J, is it I(0) or J(1)?
   Answer: 1

So in binary code the letter J is represented by the string 001011. Six "bits" of information specify one letter of the alphabet, and 20,000 million bits of information in one chromosome are equivalent to more than 3,000 million letters of the alphabet. Printers tell us that there are, on the average, about six letters in every word,* so that one chromosome may contain the information equivalent of 500 million words; with a book like the one you are now reading, having about 300 words on a page, one chromosome is equivalent to 4,000 books each 500 pages long. This, it seems, is what it takes to describe the construction, care, and maintenance of a human body. Single-celled bacteria need less information and have shorter DNA libraries; there is therefore less chance of a copying mistake when the DNA is replicated, and they are slow to evolve. Once complex creatures with long chromosomal DNA molecules appeared on Earth, copying mistakes—evolution—became more likely, and this is precisely what we find in the fossil record of life on Earth, although there are many hiccups in the evolutionary path that have not been fully explained.

A living organism such as a human being is built from the forty-

---

* This is true of "ordinary" English. The magazine *New Scientist*, which I often contribute to, does not fit this pattern, and the printers of the magazine have to allow in their planning for the fact that the words used by the contributors tend to be significantly longer than the "normal" six letters they allow for in other magazines published by the IPC group!

six–volume DNA blueprint contained in one original cell. For human beings and other life forms which reproduce sexually, it happens that the special initial cell is produced by the fusion of two cells, one from each parent, which each contain only twenty-three volumes of the blueprint, and this is why offspring inherit some characteristics from each side of the family. The forty-six volumes, in fact, consist of twenty-three *pairs* of volumes which are, for practical purposes, interchangeable, but from which only one gene, in each pair, is actually used (say, to decide the eye color, or length of fingers, in the new individual); this is very important for evolution, but once a fertilized, single human cell begins to develop, the original plans are faithfully copied each time the cell divides (a process called mitosis) so that every one of the thousand million million cells in my body, and in yours, contains a perfect replica of the original plans for the whole body.* This is the basis of the idea of "cloning"—taking a single cell from a living creature and causing it to develop into a genetically identical replica of the "parent." But this is getting way ahead of the story of the *origin* of species, and with minds suitably boggled at the complexity of some forms of life today, we must return to the slightly postprimeval soup where our first known ancestors lived.

The cell remains the basic unit of life as we know it, for man or bacterium. We don't know how the first replicators evolved, or how they came to "invent" cells. But we know that cellular living organisms existed on Earth more than 3,000 million years ago, and we can explain at least the broad outlines of how their descendants developed and evolved up to the present day. The selection of cells from one generation to the next depended on how well suited they were to exploit their environment at the time, and this indeed is what lies behind Darwin's concept of "survival of the fittest." This does *not* refer to "fitness" in the athletic sense, but simply to the fact that cells (or more complex creatures) which are fitted to their particular environmental niche will reproduce more successfully than their less "fitted" competitors. The

---

* Strictly speaking, not quite all of the thousand million million cells contain identical DNA blueprints. The cells which do the job of reproduction, sperm cells in men and egg cells in women, each have twenty-three chromosomes which have been made by shuffling the parent genes to make new combinations. This intriguing story will be unfolded shortly.

survivors today are descended from a long line of successful repro-
ducers, and "fitness" in the Darwinian sense essentially means
ability to survive and reproduce. Natural selection—a term which
Darwin used both to draw a distinction with and make a comparison with the way stock breeders artificially "select" animals for
breeding on the basis of whether they have long legs, or produce
high milk yields, or whatever—actually operates way down at the
DNA level. The animals or plants which are most successful in
evolutionary terms—most "fit" for survival—are the ones which do
a good job of passing on replicas of their chromosomal DNA to
later generations; this is the underlying concept behind the modern understanding of evolution, described neatly by Richard Dawkins in *The Selfish Gene*. The genetic (chromosomal) DNA
is not "interested" in the well-being of the body or cell it inhabits, except as a means to ensuring the spread of that particular
DNA. As Dawkins puts it, we are all "survival machines," built by
DNA and manipulated by our genes to ensure that they survive
and spread. The fact that life on Earth today is dominated by the
cell, and multicellular organisms, merely shows how successful
the cell survival machine has been at its task of spreading DNA
replicas.

SUCCESSFUL CELLS

After the appearance of the first replicator itself, the greatest invention of life has been the cell, a safe home which protects the
replicator molecules from their surroundings. Life on Earth can be
divided into two kinds, based upon two different types of cell, and
the difference between the two types is the most profound division
of all, far more important than, for example, the distinction between plants and animals. Both plants and animals are made up of
the same kind of cells, called eukaryotes. The name comes from
Greek roots, and means "true kernel," describing the most important feature of eukaryotic cells, which is the presence of an inner
cell, or nucleus, which contains the chromosomal DNA. And this
is the kind of cell from which we are made.

Most cells are tiny—perhaps a tenth or a hundredth of a millimeter in diameter—although egg cells can be quite large, and the

yolk of a hen's egg, for example, is a single cell. Although cells in a complex organism such as the human body may have many different shapes and perform many different tasks, all share certain basic features. First in importance is the membrane which surrounds the cell, its barrier against the outside world. Although only a few ten-millionths of a millimeter in thickness, the membrane controls the environment inside the cell by allowing only certain molecules to get in ("food") and certain molecules to get out ("waste products"). The membrane actively selects the molecules it allows to pass, "recognizing" them by their size and shape so that chemical messages to and from the rest of the body can also pass in and out of the cell; inside, floating in a watery fluid called the cytoplasm, there is a variety of specialized structures called organelles, which control the chemical processes by which food is converted into energy, messages are passed, and so on. The nucleus is the central controller of all this activity, which, by stretching the analogy, might be described as the brain of the cell. Its most important role is as a storehouse of information, the "library" which contains details not just of the workings of one cell, but of the whole body in which it resides, and of the cell's place in this grander scheme. Among the organelles, mitochondria do the job of converting food molecules into energy, while ribosomes are responsible for the construction of new protein molecules out of the available chemical raw materials. Plant cells, unlike animal cells, also contain structures called plastids, containing chlorophyll (and therefore also called chloroplasts), which are crucial in the process known as photosynthesis by which plants turn sunlight into energy. Lacking chloroplasts, no animal can derive energy direct from sunlight, and all animals are dependent on plants (or on other animals which have themselves eaten plants) for food. In other ways, though, animal and plant cells are basically similar—and much more like each other than they are like the cells called prokaryotes (*pro* from the Greek for "before," and meaning prenuclear).

Prokaryotic cells make up only two families among the variety of life on Earth, the bacteria and another single-celled life form sometimes called "blue-green algae" but more accurately (because of their prokaryotic family resemblance to bacteria) described by the alternative name cyanobacteria, which I shall use. The cyano-bacteria do produce oxygen as a by-product of photosynthesis, like the plants, and this has played a vital part in the history of our planet over the past few thousand million years. But both cyano-bacteria and bacteria are single-celled species which reproduce simply by splitting the two; their cells have no organized nu-clei, but merely a few strands of DNA (in the simplest, a single strand of DNA) floating in the cytoplasm within the cell mem-brane. It is obvious why biologists class prokaryotes as "pre" eu-karyotes, and this simpler form of cell is much nearer to the form we would expect to evolve from a chemical soup in which replica-tor molecules (DNA) had come into existence but were initially unprotected from the random chemical processes still going on in their environment. Indeed, it seems very likely that both mi-tochondria and chloroplasts are derivatives of what were once free-living organisms in their own right; Lynn Margulis, of Boston University, is perhaps the strongest proponent of this idea, that modern eukaryotic cells developed from a combination of pro-karyotic predecessors that learned to live together for their mutual advantage. Both mitochondria and chloroplasts contain fragments of DNA that resemble prokaryotic DNA, and the idea that eu-karyotic cells developed from a combination of prokaryotic orga-nisms is certainly no more outrageous than the fact that many millions of cells can "get together" (or, better, "grow together") in cooperating to make one animal or plant. It seems that earlier forms of life have learned to cooperate to make cells, while cells have learned to cooperate to make larger organisms—and, indeed, in some striking examples (such as bees) individual creatures have "learned" to cooperate on the next scale up, and all for the benefit of the survival of a few strands of DNA in their cells. Life can indeed be complicated, given the variety of niches open to life on Earth and the great time over which evolution by natural selection has been operating.

If the biologists are correct in their terminology, we would ex-pect to find evidence of eukaryotic life on Earth only after pro-karyotic life had become established. This is precisely what we do find; although, to understand the findings, a slight diversion into the realms of geology and paleontology (the study of fossils) is necessary.

FOSSIL HISTORY

I say "slight" diversion here not because the subjects lack complexity—far from it—but because I intend only to pay passing attention to their importance. The complexities are chiefly those of interpreting the record of Earth history preserved in the rocks and placing it in order. Once this has been done, the record tells us a great deal about how life has evolved, and what it tells us about the early stages definitely bears out the suspicion that prokaryotes did come first, as far as cellular life was concerned.

In order to understand how the history of the Earth, and of life on Earth, has unfolded, we need to know how old different geological samples are. The first stage in developing the geological "calendar" came as geologists determined which rocks are younger and which older among those available for inspection today. During the eighteenth and nineteenth centuries geologists, mainly based in Britain and Western Europe, built up their picture of the divisions of geological time, and gave rocks of particular ages names (usually in Latinized or Greek form) corresponding to the regions of Europe in which they were found—so, for example, a particular time period was given the name Cambrian, from the Roman name for Wales, where many Cambrian rocks are found. To some extent, the oldest rocks of the crust today ought to be the ones which lie deepest beneath our feet, buried by successive layers of younger rocks. But as the crustal material has been bent, broken, and distorted by the forces involved in tectonic activity, with whole mountain ranges being thrown up in some places at some times, while elsewhere and at other times crustal material is ripped apart by tectonic forces, this is hardly a reliable guide. Instead, the main guideline for establishing the order in which rock layers formed and the boundaries between different periods of geologic time (eras, periods, and epochs) depends upon the fossil remains of once-living creatures that the rock layers contain.

The boundaries are relatively easy to assign. The major division, into eras, corresponds to fundamental changes when many species in the fossil record disappear over a short space of time (a thin layer of rock) and are replaced by new species; such events are called "faunal extinctions." The lesser divisions, into periods and epochs, depend on subtler changes in the fossil record, which is why different authorities sometimes quote slightly different dates for the boundary between, say, the Triassic period and the Jurassic. In this context, the dates are best seen as educated guesses, with a margin for error of a few million years either way not being significant for a date quoted as, say, 190 million years ago. If fossils are the remains of living creatures, which they are, and fossils are important in determining the geological calendar, then clearly there is a great danger of circular arguments if the geological record is used to interpret the evolution of life on Earth. Fossil A, say, found in stratum A, is seen as biologically more advanced than fossil B in stratum B, so the A rocks are classified as younger than B. Then, perhaps a generation later, along comes a biologist studying the origins of life who muses that, since more advanced life forms are found in rocks of stratum A, which are younger than those of B, life must have evolved! Fortunately, though, this problem has been almost entirely eliminated by the twentieth-century advances in dating techniques, especially the techniques depending on measuring radioactive decay products. So although today the geological calendar still carries the names of the geological time divisions inherited from a century or more ago, the dates on those calendars are determined unambiguously by radioactive dating techniques. Where our predecessors could only get a picture of relative time (younger or older), we have the extra information provided by direct measurements of "real" time (specific dates in years, give or take a few million).

Over the long sweep of geological time, very long-lived radioactive isotopes are appropriate for use as "clocks" using the half-life technique, identical in principle to the radiocarbon calendar which is so useful for dating organic samples less than 50,000 to 100,000 years old. The isotopes of uranium—more than a dozen of them in all—are all radioactive and decay through a variety of intermediate products to lead. Uranium-238, for example, decays to lead-206 with a half-life of 4,500 million years, and uranium-235 decays to lead-207 with a half-life of 710 million years. It is a mark of just how long it is since the Solar System formed, and of just how much in the way of heavy, radioactive elements was produced in the preceding supernova explosions which processed the original interstellar material, that virtually all of the lead on Earth today is the result of radioactive decays of this kind.

The second family of radioactive geological clocks is provided by the decay of potassium-40 to argon-40 with a half-life of 1,300 million years. All of the techniques of measuring geological ages are difficult in practice, and the potassium-argon technique is made no easier by the fact that argon is a gas, and will tend to escape from a rock unless trapped in some way. But these difficulties can be overcome, and all that matters here is that accurate measurements of the ratios of lead and uranium isotopes, and of potassium and argon isotopes, in rock samples do give us a good guide to the age of those rocks, just as other, similar measurements of meteoritic samples have given us a good guide to the age of the Solar System, and even to the age of the Universe.

Armed with this information, the geological record tells us a great deal about the evolution of life on Earth. But the fossil remains preserved in the rock layers are by no means the only living forms that were alive when those rock layers were being laid down, and it is always important to remember that the fossil record is incomplete. There must be very many species—by far the majority—which have lived, reproduced, evolved, and died out over the long history of the Earth without leaving a trace for modern paleontologists to study. One reason for this is that, with rare exceptions, only hard bits are preserved as fossils, so that soft animals and plants leave few traces in the fossil record.

For animals which have skeletons—vertebrates—it is the bones of the skeleton that are most easily preserved, along with the teeth. For creatures with shells, it is the shells that become fossilized. But in both cases the remains may first be crushed, broken up, and scattered before being preserved, which makes the reconstruction task of the paleontologist that much harder. There are different fossilation processes, but in all of them (taking an oversimplified view of what happens), the hard bony or shell parts are altered by chemical reactions that replace their constituent

A fossil ammonite, typical of the enormously successful variation on the theme of life that proliferated across the oceans of the world between about 300 million and 200 million years ago. © *British Museum* (*Natural History*)

compounds by minerals from water trickling through and around the remains. The original material is eventually all gone, washed away by the water, but in its place there may be a replica in stone of the original shape, sometimes with very precise details of the original preserved. Alternatively, a shell already set in a case of permeable rock through which water trickles may be completely dissolved away, leaving a hollow "mold" in the shape of the shell; and just occasionally a soft animal such as a jellyfish or worm may die and be covered by soft mud which sets hard to preserve an outline of the animal's shape, while other imprints (footprints, worm trails, and so on) in soft mud can be preserved in the same way. And this is where we get back to the story of the spread of life on Earth.

The earliest direct evidence of life on Earth comes from just such traces of the outlines of soft creatures, colonies of bacteria and cyanobacteria, in rocks more than 3,000 million years old. But there is very little in the way of fossil evidence of life in any rocks more than about 600 million years old, and for all practical purposes the geological calendar begins with the sudden spread of a multitude of complex life forms, revealed by their fossil remains, after that time. This first and most important boundary in the geological calendar marks the beginning of the period known as the Cambrian. In geological terms, *everything* that went before, the first 4,000 million years or so (about 90 percent) of the history of the Earth as a planet, is called simply the Precambrian, and very little is known about it compared with our more detailed understanding of the past 500 or 600 million years. The same distortion, like looking through the wrong end of a telescope, appears for more recent divisions of geological time, since we know most about more recent times, for which the geological strata are easy to identify and still plentiful, whereas older strata have been broken up by erosion and other processes, and reworked into new formations. So by about 65 million years ago, the beginning of the Tertiary, we need a finer timescale than the system of periods, and the Tertiary is broken up into epochs. The perspective, once again, is that the Tertiary covers about 10 percent of the time since the start of the Cambrian; and focusing still closer to home, the current period, the Quaternary, began less than 2 million years ago, and we know a great deal more about the past 10 percent of the time since the beginning of the Tertiary (say, the past 5 or 6 million years) than about the rest of that period, and a lot more still about the past 500,000 years, or about what happened last week!

EARLIEST DAYS

But it was during the long Precambrian that life got its grip on the Earth, and evolution started life down the many-branched trail that was to lead to the diversity of species we know today. The reason for the boundary between Cambrian and Precambrian is simply the sudden, by geological standards, spread of life indicated by the diversity of fossil remains in Cambrian rocks. In other ways the rocks are no different from the rocks of the immediately preceding Precambrian. And the reason for the variety of fossils since the Cambrian is that it was at that time that living creatures had developed easily fossilized and clearly identifiable features, such as shells. There was a great deal of life before the Cambrian, but it was soft life, leaving little trace in the fossil record, and microscopic life (single-celled forms), which left only, literally, microscopic fossil remains. Along the way those "simple" life forms managed to have as dramatic an effect on the Earth as any subsequent activity of life, converting the atmosphere from a mixture of volcanic outgassing products, including plenty of carbon dioxide, into the oxygen-rich mixture of gases we have today—which then provided, as we shall see, a shield which was vital in allowing life to spread onto land, as well as the oxygen that is essential for the rapid activity of animal life. All this took time, which is why life diversified and spread so explosively only after the stage had been suitably prepared during the Precambrian. For the moment, in the interests of getting the geological record straight, I will not worry too much about *how* any living creature, including those earliest cells from which we are descended, manages to evolve. Just accepting that evolution does occur, and that more successful replicators are, by definition, the ones which have more offspring, so that any cell which accidentally develops a useful new skill will soon spread that skill among many millions of other cells, is sufficient background to the story of how those early cells did evolve during the Precambrian.

With all the interweaving threads of geology, geophysics, astronomy, paleontology, and biology involved in producing a picture of how life developed on Earth, it is inevitable that there will be some repetition in the story. But each time we come across a familiar strand we should see it from a different perspective, helping to fill out the overall picture. First, perhaps, we can narrow down slightly the limits of the period under consideration. The most scrupulously accurate modern geologists would set the end of the Precambrian as 570 million years ago, rather than the approximate "roughly 600 million years" which serves as a rule of thumb. And fossil remains of jellyfish, worms, sponges, and the like have now been identified in rocks dating back to about 100 million years before the end of the Precambrian. It is for times older still than this (more than about 700 million years ago) that the single-celled life forms can be said to have dominated the Earth. In still older rocks, there are microfossils and curious pillarlike fossils made up of layers like a stack of sheets of paper, called stromatolites. It is only in the past thirty years that paleontologists have confirmed their suspicions that stromatolites are the fossil remains of large colonies of cyanobacteria and bacteria, with *living* stromatolites identified in Shark Bay on the western coast of Australia! They are rare today because, except in special places hostile to other forms of life (the shallow waters edging Shark Bay are too salty for other forms of life), they simply get eaten; but their presence, virtually unaltered descendants of Precambrian life forms, emphasizes once again that evolution is not a process of one species replacing another. Once a good replicator system has been developed, appropriate for a particular environment, it can persist for great spans of time even by geological standards, and the spread of life on Earth is as much—more—a story of diversification into every available ecological niche as of "improvement" in the living organisms. A human being is more complex than a bacterium, but the bacterium is at least as successful as a "life machine" preserving and replicating its own genes.

So we come back to the story of prokaryotic and eukaryotic cells. The microfossil remains clearly indicate that prokaryotes

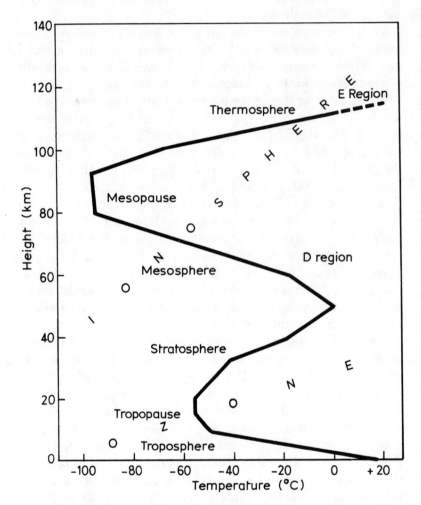

The nature of the Earth's atmosphere today is a legacy of the activity of the first single-celled organisms that learned to live with oxygen, back in the Precambrian. Because our atmosphere is rich in oxygen, photochemical reactions have led to the production of a layer rich in ozone, the triatomic form of oxygen. This is the stratosphere, which both acts as a "lid" on the weather systems of the troposphere below and provides a shield against damaging ultraviolet radiation from the Sun. The layered structure of the atmosphere is best seen in terms of temperature changes with altitude—the region where temperature increases with height indicates that energy is being absorbed in the stratosphere; this is almost entirely ultraviolet radiation, which would otherwise keep the surface of the Earth sterile.

came first, and that the larger cells typical of eukaryotes are a later arrival on the scene. But the different abilities of the two kinds of cell already reveal that this must have been the case, since eukaryotes almost all require oxygen to live (even the few odd eukaryotes who don't require oxygen today seem to have evolved from ancestors which did need oxygen) whereas prokaryotes of one form or another show a wide variety of oxygen requirements. Some bacteria cannot grow or reproduce if any oxygen is present, others tolerate oxygen but can get along quite well without it, and there are prokaryotes which don't actually need oxygen but reproduce best if there is a little around (less than the concentration in the atmosphere today) as well as some who cannot manage without oxygen. This is just the pattern that we might expect if prokaryotes had diversified into different ecological niches, forming different single-celled species, while oxygen was slowly building up in the atmosphere. And the absence of such diversity in the oxygen needs of eukaryotes then implies that they developed only after the oxygen concentration of the atmosphere had reached something near its present level. All of the evidence confirms that prokaryotes were indeed "pre" eukaryotes and remain the earliest identifiable native life forms of planet Earth.

OXYGEN AND LIFE

The evolutionary relationship between prokaryotes and eukaryotes is shown particularly well by the way they obtain energy. In eukaryotes, including you and me, the basic process by which food is turned into energy is respiration, in which glucose derived from the food (or from photosynthesis in green plants) is "burned" with oxygen to produce carbon dioxide and water, and release energy.*
Some prokaryotes, the ones that can live with and use oxygen, can also do the respiration trick, but many depend solely on the simpler process known as fermentation for their energy. In fermentation the glucose is simply broken down without any reaction involving material from outside the cell. This releases some energy but nowhere near as much, per glucose molecule, as respiration.

---

* This is slower than the burning of a coal fire, or of petrol vapor in an engine, but follows the same chemical principles.

In both processes, the energy produced is "captured" by using it to build up molecules of adenosine triphosphate (ATP), which has a high-energy chemical bond in the phosphate links. (Other phosphates can be used, but ATP is far and away the most common.) Spare energy is thrown away as heat, and when the cell needs energy it is obtained by breaking down ATP. In respiratory metabolism there are two main steps to the production of ATP. The first, called glycolysis, breaks up one molecule of glucose containing six carbon atoms into two molecules of pyruvate, each with three carbon atoms. This needs no oxygen, and produces two molecules of ATP, each with its energy store. The next step, called the citric acid cycle, takes the pyruvate and uses it, in combination with oxygen, to make thirty-four more molecules of ATP. Fermentation is very similar to the first stage of respiration—glycolysis—without this oxygen bonus; this means that for each available glucose molecule respiration produces overall eighteen times as much energy (thirty-six ATP molecules instead of two) as fermentation does. Respiration releases almost all of the chemical energy in the bonds of the complex glucose molecule, leaving only the simple molecules of water and carbon dioxide; and although the similarity between fermentation and glycolysis hints very strongly that respiration evolved from prokaryotes (as they learned to live together in ordered cells as eukaryotes), the huge energy bonus provided by respiration explains why organisms that could do the trick developed into so many species, including animals, burning energy almost profligately with their active, rapid movement, while the fermenters were stuck with a more sedentary life-style.

However, just as living stromatolites survive in a few places in the world today, even "advanced" respiration-based cells have not utterly discarded their origins. In the muscle cells of mammals, for example, oxygen deprivation can occur when prolonged activity causes a demand for more oxygen than the lungs and blood can supply, a situation known to long-distance runners as "oxygen debt." When this happens the citric acid cycle can no longer work effectively, but the cells continue to function at a reduced level of efficiency on the glycolysis method alone. "Extra" pyruvate produced as a result is carried to the liver, where energy is put in (from six ATP molecules) to convert it back into glucose. Or rather the pyruvate is first turned into lactic acid in the cell, and this lactic acid is transported to the liver and converted first into pyruvate and then into glucose. This curious diversion, which comes into play when the cells and the whole animal are under great stress, seems to echo the way bacteria dispose of waste products as lactic acid, and the oxygen-starved muscle cell is reverting, as best it can, to the metabolic processes which provided energy for our single-celled ancestors back in the Precambrian before free oxygen was available. In terms of the diversification and origin of species, and of the survival of selfish genes, the genes which enable the cell to do this trick have survived for 3,000 million years; and the reason why they have survived is that they give an advantage to a cell which can, in desperation, gain some energy when deprived of oxygen, compared with cells that have discarded these "old-fashioned" genes but suddenly find themselves short of oxygen and die. In human terms, our ancestors who had the extra "stamina" at cell level to cope with a temporary oxygen crisis survived, while our competitors without this knack died. And it does help to explain why marathon runners can keep going for so long—and why they get so hungry after a race!

Although some cells, such as muscle cells, can survive temporary oxygen debt even in organisms that need oxygen (aerobic organisms, as opposed to anaerobic ones, which can do without oxygen altogether), one all-important cell process cannot proceed, in eukaryotes, without oxygen. No eukaryotic cell can divide by mitosis unless there is at least a little oxygen present, which means that the eukaryotic single-celled organisms could not reproduce without oxygen, surely the clinching evidence that they evolved after the atmosphere had been transformed. How, then, was the atmosphere transformed—and when?

Working backward, and taking the "when" question first, we can get a good idea of the time oxygen appeared in the atmosphere of the Earth by looking at the Precambrian fossil evidence of eukaryotic remains.* Some fossil filaments similar to modern

---

* In this discussion I follow the summary provided by J. William Schopf in his article "The Evolution of the Earliest Cells," which appeared in the September 1978 issue of *Scientific American*.

fungi and green algae have been found in Siberian rocks about 725 million years old; eukaryotic microfossils from the eastern Grand Canyon have been dated as some 800 million years old, with a similar age for some Australian algae microfossils. So oxygen-breathing eukaryotic life was firmly established by 800 million years ago. Further back in time, remains that look very like eukaryotic cells have been found in rocks as much as 1,500 million years old, and these include microfossils so well preserved that it is possible to see the outline of what seem to be organelles within the cells. But although many microfossils older than 1,500 million years have been found, none of them seem to be definitely identifiable as eukaryotic. About 1,500 million years ago, there is a clear break in the microfossil record, as significant as any distinction between geological periods determined from changes in the macrofossil record. If we weren't now stuck with the traditional geological calendar, a strong case could be made for giving a new name to the period before 1,500 million years ago—the Pre-Precambrian, perhaps?—and such a boundary would indicate the importance of the break in terms of the evolution of both life and the Earth as a whole. From some time around 3,000 million years ago to about 1,500 million years ago, prokaryotes ruled the Earth. Then came eukaryotes, to be followed by all the species that could thrive on a planet with an oxygen-rich atmosphere. And the biological processes which produced the oxygen, and continue to recycle it today, must have started their work sometime in the interval between 3,000 and 1,500 million years ago. Once again, so-called primitive forms of life ("primitive" enough to have survived efficiently for 3,000 million years) still exist on Earth today, doing the same job that their (and our) ancestors must have done back when there was no free oxygen around, getting energy out of sunlight by photosynthesis. Whereas most of the photosynthesis that goes on in green plants today (and in cyanobacteria) releases oxygen to the atmosphere as an end product, some kinds of bacteria neither require oxygen for metabolism nor produce oxygen in photosynthesis—and, indeed, they cannot do the job if oxygen is present. Like the stromatolites of Sharks Bay, they seem to be survivors of a bygone age, a time when free oxygen was pure poison to everything that lived on Earth (or, rather, in the seas of the Earth, since this is long before life began to colonize the land).

PHOTOSYNTHESIS

In the first step of photosynthesis light energy is absorbed by molecules sensitive at particular wavelengths, in the exact reverse of the process by which those molecules would emit energy at distinct wavelengths, producing bright lines in the electromagnetic spectrum, if they had extra energy to get rid of. The solar energy is then used to drive a series of chemical reactions beneficial to the organism which is doing the photosynthesis. This is a complex process, which involves electrons that have been energized by the incoming radiation (sunlight) being passed along a series of molecules called electron carriers, with the resulting tiny electric current being used to break up water molecules into hydrogen (which the organism needs) and oxygen, and to convert molecules of adenosine diphosphate (ADP) into the now familiar ATP. Hydrogen, carbon dioxide, and ATP go into the next series of reactions to make glucose, while the oxygen, in modern plants, is thrown away into the atmosphere. Only the first step in the process needs sunlight, and once ATP has been made, the organism can carry on with the rest of the job in the dark, at night. The glucose is then available for the complex variety of chemical reactions involved in metabolism, either within the photosynthesizing organism or in some other organism which has eaten the photosynthesizer. Without photosynthesis, we wouldn't just be short of green plants—there would be no animals either, since animals cannot convert sunlight into energy in the form of metabolizable compounds, and we are completely dependent on plants for food, whether we eat plants directly or eat other animals that have eaten plants in order to live and grow.

But why *green* plants? The color depends on exactly which wavelengths of sunlight are being absorbed, and green represents the "leftover" radiation which is not absorbed but reflects back as visible light. Green plants use molecules of chlorophyll to do their energy absorbing, using light in the red and blue-violet parts of the spectrum and leaving most of the yellow and green sunlight to be reflected away. This is very curious, because our Sun radiates a great deal more energy in the yellow/green part of the spectrum than in the red and blue-violet, and there are other compounds which could be used in photosynthesis much more efficiently than chlorophyll. Indeed, some plants have adapted to use other pigments, reflecting away the red light and absorbing high-energy yellow/green light so that they look red. We know these plants have evolved from photosynthesizers that used chlorophyll because they too still use chlorophyll—the energy absorbed by the "new" pigments is passed *first* to chlorophyll (a completely unnecessary step) and then on down the chain to where it is needed.

Two important things—one definite, one speculative—can be learned from this. First, such a pattern, in which later evolutionary adaptations are tacked on to modified existing systems, is typical of the way evolution works. At a grosser level, in animals the organs that were fins in fishy ancestors have been adapted to become legs and arms; the land animals have not "lost" their flippers and then "invented" arms and legs. This pattern of change is particularly clear from the study of fetuses, which shows that every human being develops in the womb through stages which are very like fish, reptiles, and nonprimate mammals before becoming distinctly human in appearance. This "recapitulation" is a consequence of the fact that we are descended from ancestors of all these forms, each of which developed from an egg. The changes come by adding on to the development from the egg new, improved designs (or new, disastrously bad designs, but those don't have offspring around today!). At each stage, there is enough leeway for change in the DNA code to produce a slightly different "model" of what went before, like the difference between a 1981 car and a 1982 version of the same basic car. But the complex web of interactions that keeps a living organism alive would break down if wholesale changes were made, like ripping out the engine and sticking a steam engine in the same hole.

The second, more speculative interpretation of the curious way plants on Earth seem to be adapted for light different from the light of our Sun rests upon the argument that photosynthesis may not have been invented on Earth. If Hoyle and Wickramasinghe are correct, then even photosynthesizing cells may have developed in space before life reached Earth, and perhaps chlorophyll is a much more appropriate pigment for efficient absorption of light under those conditions. Hoyle and Wickramasinghe themselves do not go so far in their book *Lifecloud*, but point out that the chemical rings which form a basic part of the chlorophyll molecule (each ring containing one nitrogen atom and four carbon atoms, $C_4 N$) are in the porphyrin family and can exactly explain the absorption of light in interstellar space at 4430 Å. This is still a controversial interpretation of the evidence; but they say that "the basic constituents of chlorophyll may therefore well have been added to the Earth," and that would certainly suggest a good reason why life on Earth should have started photosynthesizing with the aid of chlorophyll rather than some other compound. Once it *had* started photosynthesizing, of course, competitors could never set up in business, since organic material was all rapidly getting locked up in living cells, and any improvements to the system—such as the addition of red pigment—had to come later.

Back at the beginning of the story of life on Earth, though, the first photosynthesizers differed from most of their present-day descendants in one crucial respect. They did not "throw away" the oxygen produced as a waste product of photosynthesis, but carefully packaged it up in combination with other molecules as a safe, nonreactive compound before ejecting the waste from the cell. To those first living cells, oxygen was a poison, reacting so violently and quickly with organic compounds that it would disrupt the metabolism of any cell that let it go in its pure form.

When the first cells that learned to live with free oxygen appeared, they had a huge advantage over other life forms. First, they no longer had to go through the energy-consuming business of packaging up their waste oxygen before throwing it away; then, as a bonus, the oxygen they threw away did no good at all for all the other cells around which hadn't learned to tolerate it! Once the oxygen producers appeared, they must have established themselves very quickly across the oceans of the world, and several lines of evidence show that the major transition from an oxygen-free atmosphere to one containing at least 1 percent as much oxygen as it does today occurred within a few hundred million years, about 2,000 million years ago.

One piece of geological evidence comes from deposits of the mineral uraninite (UO_2) found in some old Precambrian rocks. If there is free oxygen around, uraninite easily oxidizes to a uranium oxide containing more oxygen atoms per uranium atom, U_3O_8, so there was no free oxygen around when these deposits were laid down, and they are all older than 2,000 million years. But the best evidence comes from widespread deposits of iron oxides, found everywhere around the world, which are the main source of iron in the world today. These are known as Banded Iron Formations (BIF) and are all between 1,800 and about 2,200 million years old. The best explanation of how they came to be laid down is that in the original Earth ocean, with no oxygen around, iron was dissolved in its ferrous state. When oxygen became available, it triggered a very well-known reaction also involving iron and water, so that all of the ferrous iron in solution was converted to insoluble ferric oxides and settled in thick layers over the ocean bed. The well-known chemical reaction which involves oxygen, water, and iron, and which produces the distinctive red-brown ferric oxides, is called rusting. When oxygen appeared on Earth, all of the iron rusted. To do the job in such a short time—a few hundred million years—there must have been a steady supply of oxygen from some new source, and only aerobic photosynthesis (photosynthesis that releases oxygen) can explain the pattern of events.

Interestingly, most of the organic material that did the photosynthesizing must have been buried when it died, or its decomposition would have used up the oxygen it had produced while alive. But burial in the sediments of the ocean has always been the common fate of organic debris that has not already been eaten by other living organisms. And, of course, the level of oxygen in the atmosphere could only begin to build up after the Banded Iron Formations were deposited, since before all the iron had rusted oxygen was being trapped in rust as quickly as it was being produced. So the story of the Precambrian, and of the influence of Precambrian life on our planet, is just about complete, as far as we know it. The first photosynthesizing organisms, anaerobic pro-karyotes, were around at least 3,000 million years ago. The invention of aerobic photosynthesis a little more than 2,000 million years ago have the aerobic organisms an advantage and must have wiped out almost all of the earlier forms of life, while first rusting the entire ocean and then transforming the atmosphere (global pollution on a scale as yet undreamed of by man!). Eukaryotic cells developed about 1,500 million years ago in a stable environment rich in oxygen, after this transformation had been completed, and rapidly diversified. By about one thousand million years ago, sexual reproduction—a key invention, as we shall soon see—had been discovered, and over the next 400 million years even more diversification took place, producing many distinct multicellular species by the end of the Precambrian.

Since then, change has been rapid. This is partly because of the energy available to organisms which can use oxygen in respiration, energy which allows them to invade new ecological niches and to compete with one another (in the evolutionary as well as the everyday sense) for food. It is partly because of the invention of sexual reproduction, which allows for the spread of diversity among the gene pool of a species. And it is partly because, under the protecting umbrella of ozone in the new oxygen-rich atmosphere, life was able to take to the land, a whole new habitat with a whole new array of selection pressures operating on species.

AN ATMOSPHERIC SHIELD

Before concentrating on the biological reasons for this diversity of life on Earth today, it is therefore appropriate to look at the nature of the atmosphere we have inherited from the Precambrian, an atmosphere not just rich in oxygen but with its own layered structure, protecting us from the Sun's harsh ultraviolet radiation, which would otherwise render the land surface of our planet uninhabitable. Life changed the Precambrian environment as much as any physical process did, and we are still living off the benefits provided by our cyanobacteria ancestors 2,000 million years ago.

The structure of the atmosphere is seen most simply in terms of temperature. Heated by incoming solar energy, the ground is warm and radiates heat at infrared wavelengths back out toward space. Some of this heat is trapped by the greenhouse effect; some incoming solar radiation is reflected away by clouds, snow, and the land or sea surface. Overall a balance has been struck, although there are minor variations in the balance from time to time, sufficient to produce the pattern of repeating ice ages and warm epochs characteristic of the past few hundred million years. The lowest layer of the atmosphere is called the troposphere, and through the troposphere temperature falls off, initially by about 6°C for every kilometer of altitude gained. The decrease slows near 10 km altitude and stops near 15 km. From about 20 to 50 km temperature increases with altitude, from a minimum of about −60°C to about 0°C maximum at the top of this warming layer, which is called the stratosphere. Warming indicates that energy is being absorbed in the stratosphere, and the molecules which do the energy absorbing are those of ozone, a form of molecular oxygen with three atoms per molecule (O_3) instead of the usual two (O_2).

In the stratosphere, ozone is produced by a series of dynamically interacting chemical reactions driven by sunlight (photochemical reactions). Ordinary diatomic molecules of oxygen are split into their component atoms as they absorb ultraviolet energy from the Sun, and the efficiency with which free atoms of oxygen are produced depends on a balance between the number of molecules around to be split (more at low altitude) and the amount of ultraviolet energy available (more at high altitude). Once free oxygen atoms are produced, they combine with other diatomic molecules to make ozone, O_3. Because of the factors affecting the photochemical reaction rates, the ozone concentration is greatest in a band of the stratosphere between about 20 and 30 km altitude, and the ozone, especially in that band, also absorbs electromagnetic waves in the range 2800 Å wavelength and shorter wavelengths, stopping this ultraviolet radiation from reaching the ground. Ozone is constantly being broken up back to diatomic oxygen and single oxygen atoms, but the ozone layer is also constantly being replenished. Although the individual molecules do not stay in one state, overall a rough equilibrium is maintained, rather like a bucket with a hole in it being filled from a tap. Water is always

	GEOLOGICAL ERA	GEOLOGICAL PERIOD	EPOCH
Present		Quaternary 3 My BP	Recent 11000 y Before Present
	Cainozoic (about 1·5 percent of Earth history)	Tertiary 65 My BP	Pleistocene 3 My BP
70 MY			Pliocene 7 My BP
	Mesozoic (about 3 percent of Earth history)	Cretaceous 135 My BP	Miocene 25 My BP
		Jurassic 190 My BP	Oligocene 40 My BP
			Eocene 60 My BP
		Triassic 225 My BP	Paleocene 70 My BP
230 MY			
		Permian 285 My BP	
		Carboniferous 350 My BP	
	Paleozoic (about 8 percent of Earth history)	Devonian 400 My BP	
		Silurian 440 My BP	
		Ordovician 505 My BP	
		Cambrian 570 My BY	
about 570 MY			
	Pre-Cambrian (about 90 percent of Earth history)		

The geological divisions of time. Dates given mark the beginning of each sub-division; these have been rounded off to convenient numbers and do not necessarily add up to precisely the same figure in each column. The uncertainties involved in the dating of the divisions are greater than the differences across the columns, and it is, for example, acceptable to talk about the end of the Mesozoic as "about 65-70 million years ago," rather than to use the misleadingly precise figure 65 Myr or 70 Myr BP.

running into the leaky bucket and always running out, but the level of water inside may stay much the same.

The array of interacting processes which maintains the ozone layer is affected by changes in solar radiation, so that the concentration of ozone varies from day to night, with the cycle of the seasons, and over the sunspot cycle of solar activity. It is also affected by the presence of other chemical elements and compounds in the stratosphere, and both chlorine and some nitrogen oxides could be very efficient at shifting the equilibrium so that the overall concentration of ozone in the stratosphere is reduced—which is why there has been so much concern lately about the possible harmful effects on the stratosphere of pollution from various sources, including the nitrogen oxides released by widespread use of fertilizers, and the chlorofluorocarbon gases used in some spray cans as the "propellant" which pushes out the useful (?) product, shaving soap, deodorant, or whatever.

Above the stratosphere there is another cooling layer, the mesosphere, and at the top of this (about 80 km altitude) the temperature is a chilly − 100°C. From here on up—or out—temperature is no longer a good guide to conditions in what is left of the atmosphere, and by 500 km collisions between atoms and molecules are too rare for it even to be thought of as a continuous gas, and the outer fringes are described in terms of their electrical properties, depending on the extent to which atoms are ionized. For us, though, what matters is the troposphere in which we live, and the stratosphere immediately above which acts as a "lid" on the troposphere (because the warming layer stops convection, so that clouds and weather only occur in the troposphere below) and shields us from ultraviolet.* The crucial importance of this shield is clearly indicated by the fact that ultraviolet radiation is widely used to sterilize items such as surgical equipment in hospitals, which have to be completely free of life in the form of bacteria and the like. It is always possible to speculate about the possibility of life evolving in the presence of both oxygen and ultraviolet radiation, but the

---

* The whole circulation of the atmosphere, and the weather, must have been very different before this "lid," made of ozone, developed. Perhaps this is yet another reason why life could not get a grip on land without an oxygen atmosphere!

radiation is not just harmful because it "burns" (sunburn is produced by the ultraviolet that does get through the ozone layer). The electromagnetic energy in this wave band is particularly disruptive to DNA—the "resonance" is so strong that the molecules get broken and damaged. This is why solar ultraviolet radiation is implicated in skin cancer, which is a result of mistakes in cell division and growth produced by faulty DNA replicators. The DNA of the earliest life forms, lacking a thick skin, would have been particularly susceptible to this kind of disruption, so that life had to develop in the sea (which, in any case, is where all the organic molecules were dissolved). Perhaps a thick-shelled creature of the shallow waters, impervious to ultraviolet, might eventually have colonized the land, although it would have found nothing to eat there, but once the ozone layer was established the hazard was removed and relatively thin-skinned plants were able to spread onto land followed by the variety of animal life.

While interesting evolutionary developments continued to take place in the sea, it is from those creatures that invaded the land that we are descended, and it is their story which must be taken up in the search for the origins of mankind. Having borrowed the title of this chapter from Charles Darwin, it might seem odd that I should *end* it with the arrival of a diversity of life forms, as indicated by the fossil record, at the beginning of the Cambrian, some 600 million years ago. But Darwin himself wrote that "to the question why we do not find rich fossiliferous deposits belonging to . . . periods prior to the Cambrian system, I can give no satisfactory answer," and even went on to say that this "may be truly urged as a valid argument" against his theory of natural selection. The story of the Precambrian outlined here shows that the argument is no longer valid, and that indeed natural selection operated in the Precambrian where the origins of the species of living forms now seen on Earth do lie. The story of the past 10 percent or so of Earth history, the development of life during the Cambrian and later which was familiar to Darwin himself, is a story of explosive divergence of life, compared with the slow progress of the previous 2 or 3 thousand million years, into a bewildering variety of forms. The same biological and evolutionary "rules" operated to produce the earliest species, such as the division between eukaryotes and prokaryotes, as to produce this diversity of multicellular life. The time has come to look in more detail at those rules, and at the origins of the diversity of land-based life today.

7 THE ORIGINS OF DIVERSITY

The speed with which life diversified after about 1,500 million years ago clearly owed a great deal to the presence of oxygen in the atmosphere and the invention of respiration as a means of obtaining energy. With that background, the spread of life into different ecological niches to produce the diversity we see around us today depended to a very great extent on another biological invention,* sexual reproduction. If you look at the variety of life around you, virtually everything you can see reproduces sexually, with two different types of "parent" required to produce a new generation of "copies" of the organism. Asexual reproduction, in which one creature produces, on its own, an exact copy of itself, is almost entirely the province still of single-celled creatures like our Precambrian ancestors—although a few visible species, including greenfly and several plant species, can reproduce either sexually or asexually. Clearly, sex is a great advantage in the natural selection game, and has been for, in round terms, the past thousand

---

* Remember that "invention" is used as shorthand for "a succession of accidental favorable mutations (copying errors) which produced organisms more efficient at surviving and reproducing than their competitors." No conscious planning is involved!

million years or so. Fossil remains of the earliest known animals—jellyfish, worms, and corals—are dated around 650 to 700 million years ago, but the complexity of these multicellular organisms clearly shows that they were the products of evolutionary processes going back for hundreds of millions of years before that. So any study of the diversity of life on Earth, and of the particular evolutionary path which led to human origins, can pick up the story only about 100 million years before the end of the Precambrian. We don't know exactly how or when the first sexually reproducing multicellular ancestors of what are now land animals appeared, although we can certainly make a good guess both at how sexual reproduction developed and why it was successful. But we do know that as well as the biological factors affecting the rate of evolutionary change and the diversification of life at the end of the Precambrian and into the Cambrian, there were changes in the physical environment which must have played a part by the evolutionary pressures they produced.

In the story that follows, the geological time divisions will keep cropping up; the main division is into eras, with the Paleozoic, some 375 million years long and ending 225 million years ago, immediately following the Precambrian. The Mesozoic covers the time from 225 million years ago to 65 million years ago, and the present era, the Cainozoic, is as yet a mere 65 million years young. Within the long Paleozoic era, the subdivision is into the periods Cambrian, Ordovician, Silurian, Devonian, Mississippian, Pennsylvanian, and Permian, with the minor added confusion that the Mississippian and Pennsylvanian subdivisions were originally grouped together as one period, the Carboniferous, before studies of rocks in North America suggested the need for an extra division. To some extent, the whole Paleozoic can be divied into an early part (Cambrian, Ordovician, and Silurian) and a late part (Devonian, Mississippian, Pennsylvanian, and Permian). And this large-scale division corresponds, we now know, to physical changes taking place on the surface of the Earth, changes which were particularly important for life on Earth at the time.

The early Paleozoic corresponds to the breakup of Pangaea I, and the later Paleozoic to the regrouping of the continents into Pangaea II. As I pointed out in Chapter 4, the breakup of a supercontinent inevitably leads to a diversification of life, both because new environmental habitats are created and old ones destroyed, and because separate groups of living organisms in the shallow coastal seas are cut off from their relations in the shallow seas of what becomes a separate continent. The two groups may then follow separate evolutionary paths. In Chapter 4 I had to gloss over what I meant by evolution and diversification of life, while using the evidence of changes in the fossil record to support the idea of the breakup of Pangaea I; now, at last, I can come back to the story, having established the geophysical background, and fill it out properly by telling the tale from the point of view of life. It is worth stressing, though, that for a long period beginning rather less than a thousand million years ago, having invented multicellular bodies and sexual reproduction, life was very well placed physically. The Earth had calmed down considerably from its fiery and turbulent youth, and in the early Paleozoic there seems to have been a long period of relative calm, in geophysical terms. This followed an ice age which had affected at least part of the globe—and many people have suggested that the cold of that ice age may have been one of the factors which made the possession of a shell an evolutionary advantage, and thus led to the great spread of a diverse variety of fossilizable remains which marks the start of the Cambrian. So the Precambrian/Paleozoic boundary, chosen because of the biological changes indicated by the spread of fossil remains, "happens" to lie close to a major geophysical event, the end of an ice age and the beginning of the breakup of Pangaea I. Of course, this is no coincidence. The history of life on Earth is intimately linked with geophysical changes, and the great Paleozoic diversification took place in a world warming out of an ice age with land masses separating to provide a variety of possible habitats for life. But it was thanks to sex, as much as anything else, that life was able to exploit this particular opportunity, grasping it, we might say, with both hands.

SEXUAL SUCCESS

To understand why sexual reproduction proved such an advantage then as now, we have to look again at reproduction at the molecu-

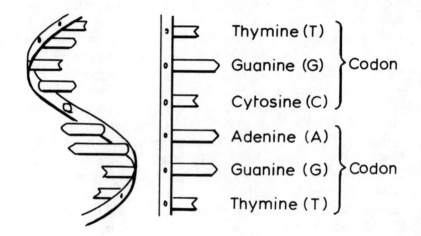

Each single strand of the life molecule DNA can be thought of as a backbone to which the side branches A, G, C, and T are attached. The full double-stranded DNA molecule is most probably a double helix, with the opposite side branches joined up to make the "rings" of a "spiral staircase." If we imagine the spiral untwisted to make a simple ladder, it is easier to see how the opposite sides join up. A can pair only with T, and G can pair only with C. When this happens, although each of the four branches is a different size, each of the two possible rungs (AT or GC) is the same size, and all of the bits of the ladder fit together.

In order to control the workings of the cell (and the body), a section of DNA separates and untwists. The broken rungs of the ladder then act as a template on which a strand of messenger RNA is built. This exactly mimics the mirror image of the DNA strand, except that a base U replaces the base T. The messenger RNA then acts as the basis for construction of amino acids, while the DNA zips itself back up.

A group of three "letters" in the DNA alphabet is called a codon, and specifies the construction of one particular amino acid.

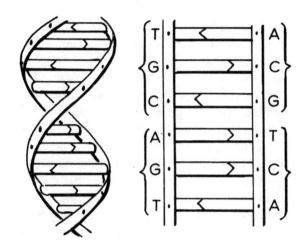

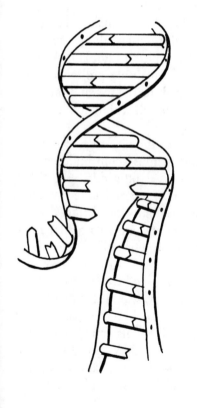

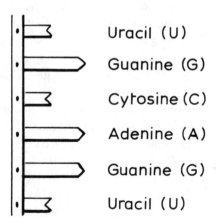

Uracil (U)

Guanine (G)

Cytosine (C)

Adenine (A)

Guanine (G)

Uracil (U)

lar level, where DNA strands are copied and passed on from one generation to another. One molecule of DNA, with its spiraling double strands matched by chemical bonds linking particular pairs of molecules, splits down the middle to produce two single strands. Then each of the two strands can quickly rebuild a whole double helix by selecting nucleotides from the biological material around it in the cell to pair up with the "broken" bonds. The four compounds which make up the four-letter DNA alphabet (A, C, G, T) only form bonds in two ways: A always with T and C always with G. So if the double helix is unzipped, one broken bond may leave A at a particular site on one molecule and T at the corresponding site on the other. The A will only recombine with another T, and the T only with another A, so that the two single strands, each opposite halves of a whole, rebuild themselves as two new DNA molecules, each identical to the whole original. It's rather as if a pair of shoes, separated and thrown into a heap of odd shoes, each paired up with another shoe to make a complete set, the left finding a new right and the right finding a new left companion. But it is rather more complex in the DNA case, where hundreds of thousands of coils, containing millions of nucleotides, have to be untwisted, unzipped, paired up, and put back together again—in the case of a bacterium, all within about twenty minutes, the time it takes for a cell to divide!

When DNA is passing messages about the workings of the cell or body it inhabits in the form of instructions about building particular protein molecules, the process is a little different. What happens then, it seems, is that only part of the molecule "unzips," leaving a free loop of DNA with the bases of its four-letter code exposed along a particular sequence, spelling out a particular message. This acts as a template for the construction of a molecule of RNA (logically enough, called "messenger" RNA) out of the chemical soup of the cell's material, the cell's own warm little pond preserved from the days of the primeval soup. RNA (ribonucleic acid) is similar to DNA but not identical, and in particular it uses one different base (U instead of T) in its four-letter code. Where the DNA spells out ATA, say, the RNA copied from the DNA template spells out UAU. But since the cell "knows" this, it causes no problems, and the completed RNA molecule is released by the DNA, which cozily twists itself back into shape, and the RNA message is then used to control the construction of the particular protein needs of the moment. Just how this happens is itself a long and fascinating story, but from the point of evolution too much of a sidetrack for even me to follow up; what matters is that the DNA does pass the messages which keep the body running, and which, even more important, control the growth of the body from a single original cell. If the DNA passed on to a next-generation cell contains a copying error, then the offspring will be different from its parent. Most often, such mutations are harmful, the offspring dies or reproduces less successfully as a result, and that is the end of the story. Occasionally, though, the mistake is an improvement, enabling the offspring to reproduce more successfully than its parent and other relations—in which case the parental strain is the one that dies out or becomes pushed aside into a specialized ecological niche where its old-fashioned ways still work to advantage, like the stromatolites in the salty seas of Sharks Bay.

We have in our cells today a great deal more DNA than our single-celled ancestors which were alive in the Precambrian. The extra material has built up from mistakes in which segments of DNA have been duplicated followed by mistakes which led to the "extra" DNA "learning" to do a useful job. In the same way chunks of DNA can be lost, so it does not necessarily follow that a cell with more DNA is more evolutionarily "advanced" than one with less. But the rules of reproduction are much the same regardless of how much, or how little, DNA the cell has, so once again it makes sense to look at modern examples to find out how evolution works, in the confident expectation that the game has been played by the same rules for as long as there has been life on Earth, and in particular for as long as there has been sexual reproduction on Earth. The DNA is arranged in chromosomes (forty-six in man, forty in the house mouse, but just in case that makes you feel superior, forty-eight in the potato), and chromosomes are made up of subunits called genes. Genes are made of DNA and carry specific messages in the four-letter A, T, C, G alphabet of DNA. Every cell in the human body carries the DNA plans which describe the building, care, and maintenance of that whole human body, although we do not yet know just how the control of production and release of certain proteins makes the difference between a single cell developing into a human being, a mouse, or a potato. One gene may carry a very simple message, such as the gene for blue eyes, but in general each gene will affect different parts of the body in different ways, and each part of the body is manufactured in accordance with a combination of instructions from many different genes. This, combined with the great many possible combinations of genes which sexual reproduction allows, is why individuals are so different from one another. For simplicity, though, we can think of genes as the basic components of the code describing how to build a body (the genetic code), and we can think of each gene, or group of genes, as carrying a simple message such as "blue eyes," "long legs," or "brown skin." Mistakes in copying genes—genetic mutations—happen all the time, since although there is enormous selection pressure for copying accuracy, there is also an enormous number of copies of each gene around today. With 4,000 million human beings on Earth, for example, virtually every gene appears in a mutated form in one human body or another, even though almost 4,000 million perfect copies of every gene are also present in the overall human "gene pool." Or, rather, not quite 4,000 million copies of each *gene*, since there may be several different versions of the gene for, say, eye color.

Each human being carries forty-six chromosomes, twenty-three inherited from each parent. So each human being has two sets of genes, and the gene for eye color inherited from the mother might say "brown" while the gene for eye color inherited from the father might say "blue." In that particular case, the human being built from that set of DNA plans won't have one brown eye and one blue; the blue gene is "recessive" and the brown "dominant," so that in practice both eyes will be brown and the gene for blue eyes is ignored. Such "competing" genes, which offer different ways to do the same particular bit of body building, are called alleles, and in the eye color example there are other possible alleles in the human gene pool, although any one individual can only have two such alleles, one inherited from each parent. Of course, both alleles may say the same thing—both instructing "blue eyes," perhaps—in which case there is no conflict. Each allele has been produced by mutation from a previous version of the gene, and it is quite possible for a large number of mutated genes—alleles—to exist in a population of a particular species. This is a key feature of the workings of evolution. Continual small changes in the genes provide for variety; then, if circumstances change so that some particular allele is favored, it will spread rapidly through the gene pool, displacing its rivals because the bodies they live in die young or fail to reproduce successfully. A new population, a variation on the previous form of the species, becomes established, and then the slow processes of evolution build up variations on the new theme, in the form of new alleles. Mutations do *not* happen suddenly, producing dramatic physical changes in the body of a new individual, compared with the bodies of its parents. Nor do mutations happen "in response to" environmental changes—the soft-bodied animals of the late Precambrian didn't "know" that it was getting colder, or that there were more predators about, and grow shells in self-defense. Rather, they must have carried an allele for thicker skin, competing with an allele for thinner skin. When the climate changed, or predators spread across the seas, the thin-skinned individuals were killed, and only the possessors of the thick-skin allele survived. Repetition of this process over many generations produced creatures with hard shells.

Taking a hypothetical human example, even though the blue-eye allele is recessive, it is widespread in the human population, and may be present even in people with brown eyes. Suppose that some change in the nature of the radiation from the Sun made blue eyes an advantage (not very likely in the protected environment of our cities, perhaps, but plausible enough if we are thinking of hunters roaming grassy plains). Then "blue-eyes" would very quickly spread—the "brown-eyes" might die of starvation because they could no longer see well enough to hunt, and making things one step more complicated, a brown-eye who had a blue-eyed son might survive if the son caught enough food for them both, while a brown-eye who had a brown-eyed son would starve along with him.

The more genetic variation there is within a population, the quicker it can adapt to changing circumstances. (Adaptation, in

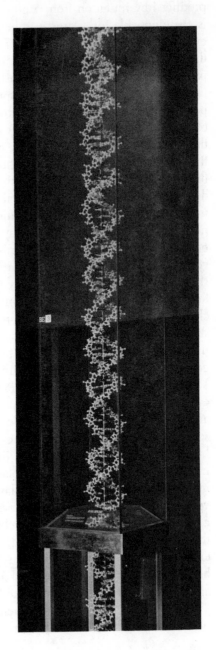

This model of the life molecule, DNA, clearly shows how the two helices are joined together by molecules, which make up the "rungs" of the spiral staircase. © *British Museum (Natural History)*

this sense, means that all the members of the population with unsuitable genes die and the survivors carry on. The survival of genes is the dominant process and driving force of evolution, and the fate of a few multicelled bodies is immaterial.) And sexual reproduction is the most efficient method known of ensuring that there is a great variety of genetic variation within a population.

When cells divide to produce a growing body, all forty-six chromosomes—or, more accurately, each of the twenty-three chromosome pairs—are copied precisely into a duplicate set of chromosomes, with one set going into each of the cells produced by the division. This process of cell division is called mitosis, and almost identical with the way a single-celled organism makes copies of its DNA before the cell splits into two—the main difference is that while those two cells then go their separate ways, in a multicelled creature the two cells stick together and help to make up a greater whole. But in the first stage of sexual reproduction a very different form of cell division, called meiosis, takes place. In meiotic cell division, whole chunks of chromosome are detached and swapped between pairs to make new chromosomes which contain the same genes as in the parent but arranged in different combinations. The cell then divides, without copying the chromosomes, in a two-stage process which produces sex cells (sperm or eggs in animals) which each contain only one set of twenty-three chromosomes.

Because of the great length of chromosomes and the amount of genetic material they contain, this process of cutting chunks out of one chromosome in a pair and swapping it with material from the corresponding paired chromosome produces an enormous variety of new chromosomes, and there is virtually no chance that any two sperm or eggs produced by one individual will carry identical chromosomal blueprints instructing how the new individual should be built. The swapping process is aptly termed "crossing-over," and the common analogy is with shuffling a very large pack of cards, although a better analogy would involve shuffling two packs of cards and then swapping cards between the two packs. And it ensures that every generation throws up new patterns of genetic arrangements from the variety present in chromosomes. Among other things, it means that although, in principle, you

could examine the material from one of your own cells and identify the twenty-three chromosomes which come from your mother
and the twenty-three from your father, similar examination of the
twenty-three chromosomes in one of your sex cells would show
that none of them could be identified as coming from one of your
parents. All twenty-three contain chunks of DNA from both of
your parents, the grandparents of the new human being that could
develop if that sex cell fused with one of the opposite sex.

This, of course, is the next stage of sexual reproduction. The organism cannot reproduce on its own, but must find a member of
the opposite sex with which to mate. When that happens, two sex
cells, each containing twenty-three chromosomes, fuse to produce
one cell with forty-six paired chromosomes, and the process of
building a new body can begin, with the specific details of each
stage of the construction being read off from one member of a pair
of chromosomes in accordance with whether the allele for that
particular stage of development is dominant or recessive on one
chromosome or the other.

There are obvious disadvantages to this method of reproduction—just finding a partner may be difficult, and the need to
search for one may make the organism susceptible to dangers that
it could avoid by hiding and reproducing asexually. In genetic
terms, from the viewpoint of Dawkins's selfish genes, it also means
that each parent contributes only half of the blueprint for the next
individual, instead of ensuring that all of its genes are passed on
intact. And all this breaking up of chromosomes and crossing-over
of DNA must introduce the chance of mistakes arising in the
copying process.

FLEXIBLE RESPONSE

But the last of these may actually be an advantage, provided there
is enough variation to ensure a flexible response to changing environmental conditions but not so much that the offspring can no
longer reproduce. And, clearly, the advantages of sexual reproduction must dramatically outweigh the disadvantages, or I wouldn't
be here writing this book and you wouldn't be there reading it.
Variability is surely the key to the success of sexual reproduction,

the variability ensured by all that crossing-over, shuffling of genes, and provision of two sets of chromosomes, with many alternative alleles, from two parents. And the circumstances in which variability is highly advantageous can be clearly seen by looking at those species of plants and animals which can reproduce either sexually or asexually—obviously, evolution over hundreds of millions of years has selected such organisms on the basis of the efficiency of their reproduction, and the survivors today must be the ones which use sexual reproduction efficiently. It turns out that the sexual reproduction stage is always associated with dispersal of the offspring into new and uncertain habitats. A classic example is provided by the behavior of parasites which may reproduce very rapidly in a population explosion within their host animal, using asexual reproduction. But when the time comes for offspring to leave the host animal and seek their fortune elsewhere, sexual reproduction comes into play, and instead of identical offspring being distributed (say, in the feces of the host animal), the new generation contains the variety that sexual reproduction makes possible. Most of the new varieties will die without finding a new host; but as long as a few of the offspring are, thanks to the reshuffling of their genes, well adapted to the host they pick up (or who picks them up) then the species continues. And, most importantly, in that case the genes which control the mode of reproduction of the parasite survive—for the genes do not "care" whether the species survives, only that they themselves are copied as accurately as possible and as often as possible.

The implication, from this and similar studies, is that sexual reproduction is the best way to spread genes in a diversity of species into a new or changing environment. That is exactly the situation which existed in the late Precambrian and early Paleozoic—the atmosphere of the Earth was changing, with oxygen concentration increasing; the climate changed, into and out of an ice age; and the continents themselves were moving, breaking up Pangaea I. Small wonder, then, that sexual reproduction received such a boost that for 600 million years or more asexual reproduction has been unable to compete. Over geological time since the Precambrian, there have been ample environmental changes to ensure that sexual reproduction continued to be advantageous in the long term. However, it is interesting to speculate that there may be no great advantage to sexual reproduction for human beings today. We have a relatively stable environment, compared with events during the Earth's past history, and we are also learning to control the environment to suit us, removing the need to adapt to changes. (If it gets cold, we turn the heating up or put on a coat, without having people prone to cold killed off to leave more hardy individuals to spread their genes for hardiness among the next generation.) The hazards of sexual reproduction might well outweigh the advantages, as far as our genes are concerned, here and now. The snag, from the point of view of the genes, is that they have built for themselves survival machines—bodies—of such complexity that asexual human reproduction has no chance of being invented: the necessary mutation would be so extreme as to make it exceedingly unlikely.*

This highlights the main problem, from the selfish gene viewpoint, of large multicellular organisms. The changes that can be made in the workings of the organism have to be subtle, and they have to be made more or less by remote control. The body is set up to do its job of reproducing and spreading genes, but the details of how and when it does the job become increasingly complex as the organism becomes complex, providing scope for the best-laid plans of genes to go astray. Even so, as long as the organism is not intelligent—and, to a surprising extent, when it *is* intelligent—the genes do very nicely for themselves, as just about every new study of the genetic basis of evolution seems to prove.

INDIVIDUAL SELECTION

In terms of its importance for our understanding of evolution, the origin of species, and the spread of diversity, the most important thing about the now thoroughly established concept of selfish

---

* And, of course, the genes responsible for sexual reproduction are quite happy with the way things are. Like all genes, they have no "wish" to be evolved out of existence, and continue to ensure their own survival by passing copies on to successive generations. Genes aren't directly interested in what is best for a species, or even what is best for the individual they inhabit. They are interested in spreading copies of themselves, and it is natural selection of the fittest in terms of survival and reproduction that does the weeding out of less "fit" individuals, and thereby of less "fit" genes.

genes is that it implies that the nitty-gritty work of selection is done at an individual level. It is an individual that lives or dies, reproduces or fails to reproduce, and it is the individual's genes that are (or are not) passed on to another individual (or individuals) in the next generation. The older idea of group selection, that individuals in a species behave in accordance with the best interests of the group as a whole, is now completely discredited. And this cannot be emphasized too strongly, since the idea of group selection has been brilliantly popularized in a series of bestselling books by Robert Ardrey. There must be a whole generation of nonscientists, or even scientists specializing in other areas of study, whose knowledge about evolution in general, and human evolution in particular, comes from reading Ardrey's books. In many ways those books are superb—certainly superbly written, and with a great deal of accurate information about past events. Where they are inaccurate is in their explanation of past events in terms of group selection, which, unfortunately, is hardly an error which can be glossed over.

So, partly thanks to Ardrey, the idea of group selection is deeply rooted in popular mythology. The still popular idea is that individuals are mere "pawns" in the struggle for survival, and can be sacrificed "for the good of the species," up to and including individual suicide if that will help the rest. And the classic example is provided by animals who give alarm calls or signals, warning of the presence of a predator at the risk of drawing the predator's attention to themselves. Surely, you may say, this has to be altruism—taking a chance on survival while letting others escape, just the kind of thing that human beings, in emergencies, get medals for? Alas, no! All of this kind of behavior can be explained in terms of the survival of genes housed within individual bodies, on the assumption that neither genes nor bodies care about the survival of the species, but only about the survival of the genes themselves. This does, however, put a slightly different meaning to the term "individual selection," since it is individual *genes*, or packages of genes, that are being selected. If one individual is killed, this need not matter to the genes if, as a result, copies of that same individual's genes survive in other bodies to be replicated. And this additional subtlety is all that experts such as Richard Dawkins and John Maynard Smith need to explain such oddities as alarm calls and seemingly suicidal behavior.

Any individual, remember, inherits half its genes from the mother and half from the father. So there is a "kinship" of 0.5 between offspring and parent. Because every offspring receives a 50-50 mix of genes from the parents, although not the same 50-50 mix in every case, there is a 50-50 chance that the siblings (brothers or sisters) of an individual have a particular gene in common—or, putting it another way, on the average siblings share half their genes. So siblings too have a kinship of 0.5. The idea can be extended outward from the immediate family: a nephew, for example, has a kinship of 0.25 with an aunt, since the aunt shares half her genes with the sibling who is the nephew's parent, and the parent shares half of its genes with the son. And first cousins have a relatedness, in these terms, of one-eighth. As far as an individual gene is concerned, inside its human or other survival machine, the important thing in life is to see copies of itself passed on to the next generation. In evolutionary terms it doesn't matter if those genes are passed on by the individual, a sibling, a cousin, or some more distant relation. And since they each share half of the individual's genes, *all* of the genes of the individual can be preserved if two siblings survive to reproduce. This is the realization that led J. B. S. Haldane to remark, more than twenty-five years ago, that he would "lay down his life for two brothers or eight cousins," a *bon mot* which should not be taken literally but which emphasizes that if a gene for some altruistic act exists—say, the warning cry gene—then it doesn't matter if the animal which gives the warning is eaten, provided that by this "altruistic" act copies of *that particular gene* are preserved in the bodies of other animals.

All this, though, applies only in a statistical sense for reasonably large numbers of animals. A particular pair of brothers, for example, might have only 40 percent of their genes in common, or 75 percent, or some other number. And in any case, other animals do not "know" who their brothers and cousins are in the same way that human animals know who their relations are. Most importantly, no animal—even a human animal—actually responds to seeing another member of the same species in danger by doing a quick calculation of how close the relationship is and then offering just the appropriate amount of help! It was Haldane, again, who commented in a scientific paper on population genetics that "on the two occasions when I have pulled possibly drowning people out of the water . . . I had no time to make such calculations." What actually happens is that, over many, many generations, evolution has selected genes which produce just the appropriate amount of altruistic behavior to maximize the spread of those particular genes. "Rash" genes, which encourage extreme acts of heroism at the slightest excuse, tend not to spread because they get less opportunity to reproduce; "timid" genes, who never give a warning to other members of the species even when desperate danger looms, tend not to spread, because the desperate danger overwhelms both the timid creature and all the nearby members of the same species, the ones most likely to be at least slightly related to the timid one. A gene for warning of the presence of danger, but not doing anything rash about it, survives best overall—if not in one body, well, that's just too bad, but all the relations run off, or fly away, to live and breed another day.

The clinching evidence that this is the basic mechanism of selection in nature comes when the numbers are put in. It is fairly straightforward to produce equations describing the behavior, in mathematical (statistical) terms, of large numbers of individuals obeying well-defined rules, whether those individuals are animals giving warning calls, or involved in mating displays, or molecules which make up a gas. Using much more complicated examples than the ones I have given, Maynard Smith, in particular, has developed models to determine just which pattern of behavior "ought" to dominate an animal society in different sets of circumstances. The branch of mathematics used by him and his colleagues is called games theory, and it has a very sound basis, thanks to the investment of effort over the years in various attempts to simulate war and predict the outcome of a chosen strategy of attack or defense before putting it to the test. At the heart of Maynard Smith's application of games theory to animal behavior is the idea of an "evolutionary stable strategy," a pattern of behavior which will persist in a population, though individuals may come and go, through many generations—because any alternative strategy which some individuals in the population might develop cannot do as well as the evolutionary stable strategy, or ESS.

STRATEGIES FOR SURVIVAL

This is particularly important because the games theory approach treats individuals as individuals doing what is best for them. There is nothing here about the good of the species; individuals act for the good of themselves. Yet what emerges, time after time, is clear evidence that species in the real world, following patterns of behavior that look at first to be tailored to the good of the species as a whole, are actually doing what is best for each individual, and are meshed in to the workings of an ESS. Dawkins, in his book *The Selfish Gene*, does full justice to these ideas and explains in detail how the ESS "works" in a variety of cases; it is hard to deny his suggestion that the concept, by showing how a collection of individuals with selfish objectives can appear to be working for the common good in line with some grander scheme of things, is the most important advance in evolutionary thinking since the time of Darwin. For this reason, and because naive acceptance of group selection ideas is still so widespread, I want to labor the point a little by paraphrasing one of Maynard Smith's classic examples of the ESS at work, quoted by Dawkins, the "Hawks vs. Doves" scenario.

Imagine a population of animals of one species, with each individual either a hawk or a dove in the sense used to denote human aggressive behavior. The hawks always fight when they meet a rival; the doves may threaten but always run away if an opponent attacks. When a hawk meets a dove, no one is hurt because the dove runs away; when a dove meets a dove, no one is hurt because after some mutual threatening they both run away; but when a hawk meets a hawk they fight until one is severely hurt. Assuming that conflicts between individuals arise over something of material value—food, or the opportunity to mate, perhaps—it is possible to allot arbitrary "points" for success or failure in the competition of individuals. This is a purely hypothetical example, so for convenience the numbers can be 50 points for a win, 0 points for running away, -100 for being severely injured, and -10 for wasting time in a mutual threat display. The points represent a direct measure of genetic success—individuals who score most get the most food and the most opportunities to reproduce, so their genes survive into the next generation. Low-scoring individuals have less chance of propagating their genes. The question which games theory can answer is: Is there an ESS in the hawks-vs.-doves scenario, and if so what is the stable balance of hawks and doves?

The first thing we learn is that a population of all doves or all hawks is *not* stable. Take doves first. If all the population are doves, then every time two come into conflict they have a threat display and each scores -10 points. But one runs away first, so the other scores 50 points by picking up the reward that was in dispute, getting a net "score" of 40. If each individual "wins" half the contests and loses half, the average score per individual per contest is 15 points, the average of 40 for a win and -10 for a loss.* Things look pretty good in an all-dove society, and nobody gets hurt, starves, or fails to reproduce. But now imagine that a genetic mutation produces one single hawk among the doves. The hawk doesn't waste time threatening, but just chases doves away, scoring 50 points every time there is a conflict. So his average score is also 50, and he is doing vastly better than all the doves, still averaging only 15 points. Hawk genes must spread rapidly through the population as a result, until there is a significant number of hawks around and they start to come into conflict.

At the other extreme, imagine an all-hawk society. Every time two of them meet they fight bitterly. One is severely injured and scores -100; the other wins and scores 50. But the average score is a measly -25, the average of 50 and -100, and any dove mutation that occurred in the population, having a comparatively better score of 0 thanks to his cowardice in running away from trouble, will do better than the hawks, at least until there are enough doves around to provide easy pickings for the remaining hawks.

Clearly, the evolutionary stable population is somewhere be-

---

* In a real life situation of this kind, there would be every incentive for one dove to increase its threat display, trying to "stare down" its rivals and win more often than half the time. This possibility is excluded from the scenario for simplicity, but it doesn't take much thought to see how this could explain the elaborate ritual displays of some animals.

tween these two extremes. For these particular points, the stable population contains five-twelfths doves and seven-twelfths hawks— seven hawks for every five doves. In the real world this is the same as saying that the stable *strategy* is for each individual to behave like a hawk seven-twelfths of the time and like a dove five-twelfths of the time, without giving any warning in advance of which course of action it would follow in any particular conflict. Genes which carry the orders, in effect, "be aggressive a bit more than half the time, but run away almost half the time" are successful and will spread to produce a stable population following a stable strategy. The fact that this has nothing to do with the good of the species is borne out by the amount of success this particular ESS produces—on the average, in the seven hawks to five doves mixed society, each individual scores 6.25 points in each conflict. This is much less than the material advantage that can be gained by each individual and by the species as a whole in an all-dove society (15 points per conflict). The *species* would do a lot better if all the individuals were doves, but this cannot happen because sooner or later one hawk will arise by mutation and be enormously successful as long as there is a majority of doves around. The aggression gene in the hawk doesn't care about the species; it cares about producing replicas of hawk genes and the success of individual bodies at making replicas of hawk genes.

Even such a simple example raises intriguing speculations about the ESS for human individuals. Do we carry "hawk" genes that are detrimental to the species as a whole in the same way? Is it possible that the relatively new evolutionary development of intelligence gives us a chance to work out the numbers and begin to act in accordance with the good of the species rather than the good of the individual, producing benefits for the individual as well, in the equivalent of an "all-dove" society? These are appropriate questions to look at in the light of the modern understanding of evolution, but they must wait a little longer, until I have covered the story of that evolution since the Cambrian, at least superficially. And before doing that, I want to offer a couple more brief examples of the selfish strategy at work.

SELFISH STRATEGIES

First, the hawks and doves brought up to date. Perhaps, long ago, there were species that were similar in size, shape, and fighting ability but differed in behavior, so that something roughly like the hawks and doves scenario really could exist. Today, of course, conflicts to the point of severe injury tend only to occur between species, so the picture is a little more complicated—but not impossibly so. To an antelope, a lion behaves like a "hawk" in our scenario, but to another lion it behaves like a "dove." This is a clear example of the instruction "sometimes be aggressive, sometimes not" being refined to "be aggressive to antelopes but not to other lions," and its value, in genetic terms, can easily be seen from the hawks and doves scenario. The antelope always act like doves where lions are concerned (and any mutation which didn't would soon loose its chance to replicate its genes!), but may follow more aggressive behavior patterns among themselves where circumstances warrant. By and large, though, over many generations the lions have become more efficient hawks—big teeth, claws, and so on—while the antelope have become more efficient doves— long legs to run away, alert senses to detect lions. Both antelope and lion, though, share a common ancestor, and one much closer to us in time than the Cambrian. Both have successfully acted as vehicles for replicating genes, and both can be regarded as successful.

A real life example of the way behavior which maximizes the chance of survival of the individual's genes appears to look like behavior designed for the benefit of the species comes from studies of the numbers of eggs laid by birds. The most powerful evidence that group selectionists were able to put forward to support their argument was the way in which many species appear to exercise reproductive restraint. Where a selfish individual might be expected to produce as many offspring as possible, spreading as many copies of the individual's genes as possible, very many species have limited numbers of offspring. This seems to fit in with the idea of group selection, since by limiting the number of offspring individuals are keeping the population within reasonable limits, so that resources, especially food, are not overexploited to the point where there is a decline in population caused by overcrowding, starvation, the spread of disease, or whatever. The classic counter to this superficially plausible argument has come from studies of birds, carried out by Oxford ornithologist David Lack and widely reported in now-standard textbooks such as Martin Daly and Margot Wilson's *Sex, Evolution, and Behavior.*

Each species of bird tends to have a characteristic clutch size—some birds lay just two or three eggs in a nest, others as many as a dozen, but with very little variation within each species. At first sight, it might look as if a small number of eggs is an inefficient way to spread genes, since a bird which lays four eggs has twice as many offspring as one which lays two. So why do some species lay so few eggs? It isn't even that they are biologically incapable of laying more, since at least some species can be encouraged to continue laying if eggs are removed from the nest—a bird which normally lays three or four eggs can be persuaded in this way to lay ten or twelve, but with only three or four in the nest at any one time. But it doesn't take much thought to realize that there must be a point of diminishing returns with increasing size of brood. After all, the parent bird has to provide food for the offspring while they are in the nest, and there must be a limit to how much food a bird of a particular species can find in a day. It is *not* true that more means better, once the maximum number that can be fed adequately has been reached. So a bird which carries a gene for producing many eggs may be at a disadvantage. In trying to feed her huge brood, the mother may not only starve them but work herself to death, so that the gene doesn't spread. At the other end of the scale, a bird which produces fewer offspring than she could rear adequately will be successful, but only in a limited way. The egg-clutch–size gene that spreads through the population will be the one which most nearly fits the maximum number of offspring that can be reared by a mother of that species in an average year.

These ideas are borne out by observations of birds in the real world. Because there is some variation within populations, ornithologists can study the success, in breeding terms, of individuals within a species which have different numbers of offspring. Although most Great Tits, for example, produce nine or ten eggs in a season, some produce only three or four, and others as many as thirteen. In an average year, observations show, the greatest number of offspring which survive to become adults come from the nests which had nine or ten nestlings. The survival of a gene—strictly, an allele—which does occasionally produce a larger number of eggs is explained because in a good year, with more food available, it does happen that there are more survivors per nest from nests in which unusually large numbers of eggs were laid. This shows, in miniature, the forces of evolution at work, as the presence of variation within the species allows it to "adapt" to slightly different conditions. If the conditions changed permanently, and the process repeated itself for a thousand generations, the alleles that were once in a minority would become common and the species would have changed into something else. By selfishly working to ensure the survival of its own genes, the individual provides not for the survival of the species, but for the survival of the best modification of the species appropriate to each particular circumstance. And the example of the importance of the parental investment in rearing the young indicates how, through the operation of selfish genes, it has come about that *two* sexes are, in most cases, necessary for sexual reproduction on Earth today.

SEXUAL STRATEGIES

Sexual reproduction without sexes sounds bizarre to us, conditioned to regard the world around us as "normal." But there is no need for two *different* types of sex cell to be involved in swapping chromosomes, and surely the first single-celled organisms that benefited from sexual reproduction were indistinguishable from one another. Sexual reproduction had to come first before sex could evolve. And although we can never know for sure just how it evolved, there is a very plausible explanation of what must have happened then.

Among the single-celled eukaryotes that had begun to reproduce not just by splitting in two but by first swapping genetic material and mixing it between two cells, each of which then split in two, there must have been a variety of sizes. Any cell could "mate" with any other, but the success of each cell at reproducing would be affected by the size of the combination produced as a result—large cells would be more successful because they contained an ample supply of material for division into a new "generation" of smaller cells still large enough to function effectively. So it might seem that natural selection would, over millions of generations, produce a shift in the characteristics of the "species" toward bigger cell size. But there is an opposed effect which also operates in the circumstances of those early eukaryotes, floating free in the ocean. Sexual reproduction can only take place if two cells happen to meet one another, so that a cell with more mobility could swim around more, having a better chance of meeting another cell and reproducing, to pass on, among other things, its genes for improved mobility. In essence, more mobility means smaller. So two kinds of cell would do best at replicating their genes once sexual reproduction had been invented: the big cells with plenty of organic material and the small cells that could swim about to find partners. For each small cell, it would be particularly important to combine with a large cell, to compensate for its own small mass, and evolution would very quickly select out any small cells which "mated" together. The large cells would still prefer to mate with other large cells, but would have less and less choice as their size increased and the small cells became ever smaller and more efficient at seeking out large companions. So two different sex cells must develop from the initial population which had a random distribution of cell sizes. The pattern remains to the present day, with the large cells (eggs) being uniquely characteristic of females of all species and the small sex cells (sperm) being characteristic of males. Regardless of body size, the male of the species produces sperm which are much smaller than the eggs produced by the female of the same species, and this is the best and simplest definition of the two sexes.

The almost equal numbers of each sex for a particular species then follow automatically as an evolutionarily stable strategy. Talking now about large, multicellular animals, each individual has to find a mate of the opposite sex in order to reproduce. If there are more females than males, say, then males immediately have an advantage, and have a better chance of finding a mate and passing on genetic material, including the allele for maleness.* So the proportion of males will increase. Similarly, if there are more males, then females find it easy to locate a mate and reproduce, so the allele for femaleness spreads in the gene pool. Anytime the balance of the sexes tips away from the fifty-fifty ratio, the evolutionary pressure acts to restore the balance. So the success of sexual reproduction as a means of mixing genetic material, ensuring sufficient variability in a population to cope, by adapting, to a changing environment, led directly to the evolution of two sexes and to the pattern in which there are roughly equal numbers of males and females. The mother's bigger "investment" in the large egg has also led to the evolution of most of the characteristics by which we distinguish sexes, including maternal protectiveness.

A male animal can produce enormous numbers of sperm cells, and "expects" most of them to die without meeting an egg and producing offspring. A female, on the other hand, produces relatively few but larger eggs, each requiring an "investment" in the form of food used in their production, so that each is more valuable to her than an individual sperm is to a male. From this tiny initial difference, selection has operated over many, many generations to widen the gap, just as the difference between egg and sperm cells originally came about. The more a mother "invests" in the survival of her offspring, the more "maternal" she is and the more successful in reproductive terms. And the more a male can get away with spreading his sperm among a large number of females, each of which pays suitable maternal attention to the resulting offspring, the more successful he will be in reproductive terms. It is no coincidence that many animal species operate on the harem principle, with a dominant male running as large a harem of females as he can keep in the face of competition from other males. And this kind of basic evolutionary development must also have relevance for human behavior, however much we may try to be "civilized" and live according to different rules from those of the most stable evolutionary strategy.†

---

* The "gene" for sex is actually a whole chromosome, indicating its importance in evolution.
† In fact, as Dawkins points out, the best strategy for *either* sex is to abandon the offspring, provided there is a good chance that the other parent will look after them. Because fertiliza-

EVOLUTION OF DIVERSITY

So evolution proceeds by the workings of sexual reproduction, and has done so for hundreds of millions of years. In looking at the story of how multicellular organisms evolved, and in particular at human origins, sexual reproduction is the only kind that matters. This method, combined with the diversity of gene alleles available in the whole population of a species, is what makes evolution at the speed which has occurred since the Cambrian possible, with new "hands" of genetic material constantly being dealt from the reshuffled "pack," and minor mutations occurring to remain in the population, perhaps at a low level, until some change in the world outside suddenly gives them high survival value and allows them to spread while others decline. Remember that a sexually reproducing animal, such as man, has two sets of chromosomes, one from each parent, and at a specific point on one chromosome will be a specific gene, perhaps the one which determines eye color. If each of the pair of chromosomes has the same gene at that point, it is homozygous for that gene. If it has different alleles at that point, it is heterozygous. Most of the gene locations are, in fact, homozygous, with the blueprint inherited from each parent describing much the same structure overall (two legs, a heart which works like this, a liver built according to that plan, and so on). Only 6.7 percent of the sites are heterozygous in man, 6,700 out of an estimated 100,000 gene loci. With chromosomes coming in pairs, this means that there are potentially $2^{6,700}$ different ways to make a human being within the overall framework. This isn't even an astronomical number—it corresponds to rather more

tion of the egg takes place inside a female animal's body, it is rather difficult for her to carry out this strategy, and the male has ample opportunity to run off with someone else. This method of reproduction is a consequence of living on dry land, and some fish confirm the reversibility of the strategy under appropriate circumstances. In the sea, fertilization of the eggs can take place outside the body, and the female must lay her eggs first before the male releases the free swimming sperm cells. Indeed, the male daren't release the sperm too soon, or they will wash away in the sea; he *has* to wait until the eggs are laid and he can release the sperm where they will do most good, in reproductive terms. By the time he has done this, the female can make her getaway, leaving the male to look after the fertilized eggs. This is exactly what happens in at least one species. And whichever way around the strategy works, the partner left holding the baby cannot abandon it, or the whole reproduction process has failed. Single parents who abandon their babies are very effectively selected against by evolution, but philanderers are not!

than $10^{2,000}$, and if we assumed that the Universe contained just enough matter to be closed, all of it in the form of hydrogen atoms, then the number of atoms in the whole Universe would be 10^{80}, absolutely tiny and insignificant compared with $10^{2,000}$! The odds against two people being genetically identical, unless they are identical twins produced by the division of one fertilized egg cell, are $10^{2,000}$ to 1, and this is the variability which provides the pool of slightly different variations from which evolution selects the fittest in each generation. If the climate changes, or a new predator appears on the scene, it is very unlikely that every individual from such a diverse population will be wiped out, and the survivors must, by definition, be the ones best fitted to survive and reproduce in the changed situation.

So evolution is going on all the time, as new permutations of alleles are shuffled into the available gene loci and the most efficient combinations survive most effectively. This gradual adaptation of a species to its (perhaps changing) environment is called phyletic evolution. Species can also diversify, splitting into two or more separate species, if different populations are separated by a physical barrier—a new mountain range, say, or the cracking of a supercontinent into smaller continents. When this happens, each branch of the family will continue with its phyletic evolution. Maybe, before the distant relations have changed too much, the split populations will be brought back together, interbreed, and become indistinguishable once again. Or perhaps they may be separated for so long, and change so much, that they could no longer breed with one another even if they did come back into contact. This is how new species begin, and the process is called speciation. Going back into the mists of the Precambrian, every living thing on Earth has been produced from repeated speciation and phyletic evolution, starting out from a very few—perhaps only one—original living cells. Multicellular organisms, however, have evolved on separate and independent occasions from several different immediate single-celled ancestors.

When we trace the ancestry of man forward from the earliest cells on Earth to the present day, the story inevitably unfolds as if evolution were working to produce a specific end product that is better than what went before. To many people, mankind is still seen as the "end point of evolution," a "superior" creature compared with all the other products of evolution. But evolution has not finished, and there is no way that we represent its end point; nor are we superior, by any biological or evolutionary standard, to other species—just different. Intelligence is a very significant, interesting, and important difference, certainly. But it seems quite possible that intelligence may directly contribute to the end of the human race, through warfare, after a very short span of time on Earth. If that happens, then far from representing the (or even a) pinnacle of evolution, we will have represented a blind alley, less successful at preserving and replicating genes than the dinosaurs, the trilobites of the Cambrian, or the prokaryotes which have survived, as types, from the Precambrian to the present day. It is important to keep in mind the diversity of life on Earth, and that every speciation which has taken evolution one step closer to producing modern man has also involved the splitting off of at least one other species from the original stock. As human beings, we are especially interested in human origins, and in a book of this size there is no space to follow every branch of the tree of speciation and diversity. But they are there, nonetheless.

So, with this in mind, we can bring the story of human origins forward from the earliest cells to the arrival of recognizable prehumans. We do not follow the story of prokaryotes, since our own ancestors are descended from the eukaryotic branch of life; we do not follow in detail the diversification of plants, since we are animals; we are less concerned about life in the sea than on land, since we live on land; and we are only marginally interested in the literally spineless invertebrates, since we all have spinal cords and bony skeletons as a framework for the soft parts of our bodies to build upon. Against this background, however, is the knowledge that there are more than 2 million multicellular species in the world today, and that many more have come and gone over the sweep of geological time since the Cambrian. Life is very complicated!

It is easy to see, for a start, why multicellular organisms are successful in evolutionary terms. Because there are many cells working together, the death or damage of one cell, or even of a lot of cells, need not be important to the organism as a whole, since individual cells can be replaced. This gives the organism a chance to live longer, and to produce many offspring, using cells which, protected by the rest of the organism, can become highly specialized at reproduction and very good at the job of shuffling and replicating chromosomes. Everything else—strength, sharp teeth, good eyesight, and intelligence included—has evolved because it helps the organism to survive and reproduce. Multicellular organisms today are all members of one of three "kingdoms," the *plants*, which make living organic matter using the energy of sunlight in photosynthesis; *fungi*, which are most simply (if inaccurately) described as plants which feed off organic remains rather than using photosynthesis; and *animals*, which feed off other organic things and also run about.*

Fossilized burrows have been found in rocks as old as 700 million years, showing that animal life was well established late in the Precambrian.† Remains of jellyfish from the time around the Precambrian/Cambrian boundary have been found in Australia, and the hard shells of invertebrates such as trilobites have left their mark right through the Cambrian, defining the boundary itself. By the end of the Cambrian, some 500 million years ago, Pangaea I was breaking up into separate continents which drifted away from one another, creating a diversity of new environmental conditions in the shallow seas around their coastlines, and leading to a great diversification of species. Among this diversity of life were our direct ancestors, the first vertebrate fishes. Unfortunately, however, the fossil record has not been laid down with the conve-

* Some animals do not "run about" in the usual sense. The sponges are actually colonies of small animals working together, and sponges represent a kind of multicellular cooperation invented by eukaryotes quite separately from the invention of multicellular cooperation which led to the rest of animal life. Sticking to my brief of following the trial of *human* origins, I will say no more about them.
† While this chapter was being prepared for the printers, Dr. William Schopf and colleagues at the University of California, Los Angeles, announced the identification of microfossil remains of five separate types of organisms, all of them looking like modern bacteria, in rocks from Australia dated as 3,560,000 years old. These are the oldest identified fossils. Although far removed from events of the Cambrian, and the evolution of life on land, the existence of such life forms from so long ago is certainly worthy of mention. And proponents of the idea that life originated in space and then seeded the Earth have been quick to point out that it is very difficult to see how such relatively biochemically advanced organisms would have evolved from scratch within 1,000 million years of the formation of the Earth. The microfossil identifications back up the evidence from stromatolites from the same region, mentioned in Chapter 6.

nience of human investigators in mind, but in accordance with the workings of wind, waves, and weather. And one of the most annoying gaps in the record, from the human point of view, comes just here. The Ordovician is shown from other remains to have been a period of major development in marine life, and fossil remains of invertebrates abound. But apart from a few scraps of bone and scaly armor plate, no evidence of vertebrate life in the Ordovician has yet been uncovered. The scraps that have been found are sufficiently like later remains for paleontologists to be sure that small vertebrate armored fish swam in the seas more than 450 million years ago. But the gap in the record has made it impossible to identify the direct invertebrate ancestors of these first vertebrates, which had no jaws and are called ostracoderms.

Although speculation about those immediate ancestors is no real substitute for fossil evidence, there is one good explanation of why there are no fossil remains of the first vertebrates. Most vertebrates today have a bony spinal column, which serves the dual function of protecting the important nerves within and providing support for the body and head. Some species around today, most notably the sharks, have spines made out of soft cartilage rather than hard bone, which is no disadvantage to a shark, supported by the sea. It seems very likely that the earliest vertebrate fishes "invented" soft, cartilaginous vertebrae first, and that the hard, bony vertebrae were a later evolutionary development.*

FROM FISH TO AMPHIBIAN

Whatever their direct origins, there is no speculation about the success of the bony fish once they did appear. There are more bony fish species in the world today than there are species of all other vertebrates put together, another reminder that our view of the human "success story" is a result of a rather special point of view. But modern species are very different from those first ostracoderms in two important respects.

* The sharks, by the way, are among the great evolutionary success stories. They have never dominated their environment in huge numbers, but have been around since the Devonian with only minor modifications. The only hard parts of a shark's skeleton, for example, are the teeth; they have never been under any evolutionary pressure to evolve a hard skeleton because they are doing very nicely without it, and have done so for about 400 million years.

As well as having no jaws, those early bony fish lacked the paired fins which make modern fish such good swimmers. A few surviving descendants of this line are around today (the lampfrey and the hagfish), but the value of jaws with sharp teeth, and fins paired up for improved swimming, is so large in evolutionary terms that ostracoderms were almost entirely replaced in the Devonian by placoderms, with their more efficient eating mechanism which, of course, implied a longer life and more opportunities to reproduce. The evolution of jawed fish from the Devonian followed two main lines, one leading to the variety of bony fish we know today and the other to the cartilaginous sharks and their relatives, the rays, which make the most of the flexibility provided by their cartilaginous skeletons. The next evolutionary diversification, however, brings us to a point where we leave the main line of successful evolution of bony fish, which becomes a story of ever-improving adaptation to the watery habitat leading to the great diversity of modern fish today. Instead we turn to the adventures of a species which became split off from the main group and has since followed a much stranger and, to us, more interesting path.

Almost all modern fish have descended from a type of jawed fish known as the ray fins, which have the streamlined fins of fish, such as cod or salmon, known now. The other major type of jawed fish, in the late Silurian and early Devonian, was the lobe fins, which had bits of bony material linked to the fish's skeleton and extending into the fins. These may have been useful for groveling around in the mud at the bottom of the water, but ray fins proved much superior for swimming, and by fish standards the lobe fins proved something of a failure, surviving today only in a few forms. But those bony lobe fins were crucial in allowing another development to take place—the evolution of limbs which enabled some of the descendants of the early lobe-finned fish to crawl out of the water and try for a new life on land. We are not only descended from fish, but from rather unsuccessful fish at that!*

* Not really so surprising, since success, as far as our genes are concerned, means *accurate* replication, and the most successful Devonian fish have fishy descendants swimming the seas today. The lobe fins, which competed directly with ray fins, were wiped out by the competition; the few freaks which spent more time groveling in the mud of shallow

No animals could make the next step, moving onto land to live, until there was something to eat on the land. In other words, plants had to colonize the land first. The best evidence we have is that there was no significant colonization of the land by plants until halfway through the Paleozoic, in the late Silurian. Plants, of course, were subject to the same evolutionary pressures as animals, and it is no coincidence that plants were diversifying and spreading into new niches at the same time that animals were diversifying. Both were the multicellular products of earlier single-celled organisms; both depended on the increasing level of oxygen in the air for respiration; and both were affected by the ebb and flow of ice ages and the breakup and rearrangement of continents. The combination of suitable factors from the Silurian onward saw the plants moving very rapidly onto the land, and fossil remains show that by the middle of the Devonian there were real forests, with plants up to ten meters high and a thick underbrush of smaller plants around them. The Carboniferous saw an even greater explosion of plant life, at a time of warm, moist climate over much of the world's land surface. The name Carboniferous comes from the extensive coal deposits laid down at that time and still being exploited; the coal was produced from the remains of huge trees, which were by then diversifying into many forms and had invented seeds. Enormous amounts of once-living debris were deposited at this time, being buried and eventually transformed by geological pressure into coal. This must have taken a great deal of carbon dioxide out of the atmosphere, releasing oxygen and locking up the carbon as coal deposits, and such a change may have influenced the further development of life, perhaps by reducing the temperature slightly as the greenhouse effect was weakened and quite probably by enriching the oxygen content of the atmosphere and so paving the way for active, oxygen-breathing animals to move onto the land. Perhaps the greatest problem of human impact on the environment today is the way in which we are rapidly burning fossil fuel, in the form of coal deposits laid down long ago, and putting carbon dioxide back into the atmosphere. There

coasts, rivers, or pools had no competition and survived, reproducing to pass on their genes for mud crawling in shallow water, and with sturdier lobe fins now being selected naturally as an advantage, allowing a fish to obtain more food and reproduce successfully.

is serious concern that this may lead to an unpleasant global warming, with detrimental effects on human life; in global terms we may be turning the clock back, in one respect, to the Carboniferous! The fact that life survived so well in the Carboniferous and afterward rather gives the lie, though, to those alarmists who foresee a *runaway* greenhouse effect as a result of burning fossil fuel, with the Earth becoming a scorched desert. Although conditions have changed in many ways in the past 400 million years, there is no real reason to suppose that the amount of carbon dioxide which proved so pleasant for life then would not be tolerable today, and there is no way we can put more carbon dioxide than that into the atmosphere, since all we are doing is restoring what was once taken away by the lush plant life of the coal-forming periods.

As plant life moved over the land, first sticking near water, then evolving techniques for survival further away (techniques such as root systems to seek out water below ground and surface coverings to stop water evaporating too easily from the plant itself), the animals followed. Worms squeezing through the mud and many-legged "millipedes" were among the first, their bodies needing little adaptation—apart, of course, from "learning" to breathe oxygen from the air instead of from water—to a land-based life, and the organic remains of plants, or living plants themselves, providing their food. Reproduction posed new problems on land, but again, nothing that couldn't be overcome with millions of years, and millions of generations, for evolution to encourage diversification and specialization. Many of the relations of these creepers and crawlers survive today, and their best-known offspring are the insects, which began to appear about the time that specialized descendants of lobe-finned fish also moved onto the land, where by now there was, at least in the wetter regions, an abundant supply of crawling things, as well as plants, for them to feed on. This new selection pressure operating on the crawlers may have been one reason why so many of them developed flight. However, although a strong argument can be made that insects have been even more successful than vertebrates in their conquest of a diversity of ecological niches on land, we are vertebrates and the story of human origins lies with those first vertebrate land animals rather than with the history of insects, except where the insects provided a valuable source of food.

So we can return to the story of the evolution of our direct ancestors, picking it up in the late Devonian, about 350 million years ago. At about this time, some lobe-finned fish had learned to breathe air as an extra source of oxygen and to use their stumpy fins to help their movement across shallow pools. Natural selection favored the offspring of those fish that had more efficient lungs and more efficient "legs," almost certainly because of the untapped supply of food available on the land surrounding the shallow pools in which they lived. Unlike the case of the missing record of the ancestors of the first vertebrate fish, we do have fossil specimens of what seems to have been the direct ancestor of this evolutionary line, from which all land vertebrates today are derived. Clearly, once one line had become established on land, latecomers had very little chance of competing with them, so no other vertebrate line was able to make the transition from water to land. So the fish called eusthenopteron, found in fossilized form in deposits 350 million, or a little more, years old is one of the clearest examples of a common ancestor, from which we have descended, along with all the other land vertebrates including elephants, my cat, your dog, a hummingbird, and an Australian kangaroo.

Eusthenopteron's special distinguishing characteristic is the passage which links its nostrils with the roof of its mouth, a unique feature among the various candidates (such as coelacanths and various Devonian lungfish) for ancestors of the amphibians, but a feature found in every land vertebrate today. In addition, the pattern of bones in the fins of the lobe-finned eusthenopteron is very similar to the pattern of bones found in all land vertebrates—the bones may be bigger or smaller in proportion to one another, but the same pattern of bones in the same relationship can be traced back from all land vertebrates to eusthenopteron. For a short time, eusthenopteron's immediate descendants had almost ideal conditions as they learned to lumber out of the water, first in brief forays, then in longer and longer expeditions in succeeding generations as the more successful were selected. To start with, there were no predators chasing them on land; then there was plenty of food in the form of worms, snails, and the ancestors of insects. So the amphibians diversified dramatically, changing this cozy picture considerably as some lines became predators themselves, feeding off their cousins. They still had to return to the water from time to time, some to keep their skins damp and all of them to reproduce—this is the key distinguishing feature of surviving amphibian species such as frogs and newts today. But modern amphibians are nothing like their ancestors which dominated the land for 75 million years, with some species several meters long and clearly flesh eaters, as indicated by the fossil remains of their sharp teeth and strong jaws.

THE DAY OF THE DINOSAUR

By now, to give some idea of the timescale over which these evolutionary changes took place, the continents were regrouping into Pangaea II, with widespread changes in the environment in many parts of the world. From the beginning of the Cambrian to the Mississippian spread of the amphibians represents a span of about 250 million years, long enough for major changes to have taken place in the geography of the Earth, as well as in the forms of life which inhabited the Earth. Even so, the development of amphibians, land based and only returning to the water for reproduction, with a larval stage like the frog's tadpole swimming in the water until maturity brought with it the legs and air-breathing mechanisms necessary for a life on land, happened with dramatic speed compared with the age of our planet. From the earliest vertebrate fish to the diversification of amphibians across the land took only a couple of hundred million years, setting the origin of vertebrates somewhere in the gap in the fossil record, about 500 million years ago. Since then the pace of evolution has been equally startling in some respects—from amphibian to man in only another 300 million years—although successful and well-adapted species (including sharks, dragonflies, and roaches) have stayed much the same ever since the late Paleozoic.

While the insects were already consolidating their positions on land, however, the vertebrates were still diversifying and throwing up new species. In the process the continual shuffling of genes brought about another change in the late Paleozoic, as significant as the change from vertebrate fish to amphibian. The great draw-back to amphibian life is, of course, the still-present commitment to the water. First, the larval form has to go through all the dangers of life in the water before escaping to cope with the quite different dangers of life on land. Secondly, the adult form of the creature cannot range with complete freedom across the land, but must live within range of water. When natural selection produced species that had developed fertilization of the egg within the mother's body and an egg that was not laid until it had been coated with a hard, protective shell, those species were at as much of an advantage over the amphibians as the early amphibians had been compared with their immediate fishy ancestors. These egg-laying animals were (and are) the reptiles; in essence, their eggs allowed their young to develop through the "tadpole" stage safe in a tiny watery world of their own, only emerging into the world outside when they were mature enough to breathe air and run about on their own four feet. The reptiles, from their origin in the Pennsylvanian, swiftly spread and replaced the amphibians in most ecological niches, some becoming grazers while others became predators, some very large—the size of cattle—and others very small—the size of a mouse. By the late Permian, reptiles were well on the way to dominance, with a great diversity of reptile species becoming established. They were, indeed, well placed to survive changes that were then taking place on Earth, and to diversify still further in the changed conditions of the next geological era, the Mesozoic.

And at the end of the Permian, things changed dramatically.* One reason for this was the tectonic activity associated with the final stages of the construction of Pangaea II. This saw great mountain ranges being pushed up, and the disappearance of many shallow seas and of low-lying coastal regions ideal for life, with plenty of rainfall from the sea breezes. In the middle of the Per-mian, about 250 million years ago, there was a major ice age

---

* This, of course, is why we label it the end of the Permian, and also the end of the greater division of geological time, the Paleozoic era. More accurately I should say "then things changed dramatically, and as a result geologists defined the time of dramatic change as the end of the Paleozoic." But you know what I mean—I hope!

which may have lasted for 20 million years, a direct result of the geography of the time and the extension of a large part of Pangaea II, including much of what we know today as India, Australia, South America, southern Africa, and Antarctica, across the South Polar region. (The traces left by the ice cover of this period, indeed, provided the first real evidence in support of the idea of continental drift.) There had been other ice ages before, and other episodes of mountain building and so on. But the particular combination of dramatic changes in the environment just over 200 million years ago seems to have been too much for many species, especially those living in the shallow seas, just as similarly dramatic changes associated with the formation and breakup of Pangaea I, some 600 million years ago, helped to give life a boost by first producing harsh conditions which brought death to many species, then easier conditions which allowed the survivors to diversify and radiate out to fill the empty ecological niches.* The same thing happened at the end of the Paleozoic, with the difference that by now life was firmly established on land, so that the diversification occurred among well-established land animals, as well as among the creatures of the shallow seas (and the plants, of course).

It is certainly encouraging, from the point of view of confidence in our understanding of the history of the Earth, that the break in the fossil record which was so important to geologists more than a century ago that they made it a boundary between two eras coincides so well with a tectonic event of major importance, revealed by late twentieth-century studies, the breakup of Pangaea II. The time from 225 million years ago to 65 million years ago is known as the Mesozoic ("middle") era, and it begins with the split of Pangaea II. From one supercontinent, across which life could

---

* Some people argue that the combination of mountain building, earthquakes, volcanic activity, an ice age, and the disappearance of shallow coastal seas is still not enough to account for the dramatic extinctions of many species at the end of the Paleozoic. They suggest that some extraterrestrial influence may also have played a part—a change in the Sun's activity or the explosion of a supernova nearby, sending sleets of cosmic rays across the Earth, for example. In my view, at least for the late Paleozoic extinctions, this is overkill with a vengeance, and it seems to me that the terrestrial changes at the time were more than enough to produce the stress on life forms needed to account for the extinctions. However, the cosmic catastrophe idea may be relevant to other extinctions in the fossil record, as we shall see in the next chapter.

spread in direct competition and contact with all other land-based life on Earth, the land mass broke up into separate continents on each of which evolution by natural selection could operate almost in isolation from the rest of the world. So once again life forms diversified, following slightly different paths on different continents, until they became many new species. One group of reptiles was particularly successful in spreading into many ecological niches as conditions for life improved after the extinctions of the late Paleozoic; these were the dinosaurs, and the story of animal life in the Mesozoic, all 160-odd million years of it, is essentially the story of the dinosaurs.

By now, a little punch-drunk from the literally astronomical numbers that have been appearing regularly throughout this chapter and the whole book, you might think that 225 million years ago was just yesterday on the geological timescale, and so indeed it was, compared with the 4,500 million-year history of the Earth. So it may be appropriate to pause for breath and take fresh stock of the situation, with the information that at the beginning of the Mesozoic there were 385 days in a year. The reason? Quite simply, for hundreds of millions of years the spin of the Earth has been slowing down gradually, thanks to the tidal influence of the Moon. The strength of the effect can be calculated, and the fossil remains of creatures which show regular growth cycles (monthly, seasonal, and so on) confirm the accuracy of the calculations. The day used to be much less than twenty-four hours long, with correspondingly more days in the year. No one, as far as I know, has satisfactorily incorporated this outside influence in any explanation of the evolution of life on Earth, although anyone with a fertile imagination can make his or her own speculations about how a twelve-hour-long day might have affected life long ago! But here I want only to point out that over the span of the Mesozoic, just yesterday by some standards, the number of days in a year declined from 385 to 371, as each day got slightly longer thanks to the insistent tugging of tidal forces. The dinosaurs were the dominant form of land-based life for so long that they saw the length of the day increase by about 4 percent and the number of days in the year decline by fourteen as a result. Our own stay on Earth is, as yet, far from being in that league, and this does give another indication of just how long natural selection has—how many generations it can operate on—to produce changes during a geological era.

The subdivisions within the Mesozoic are the Triassic period (from 225 million years ago to 180 million years ago), the Jurassic (180 to 135 million years ago), and the Cretaceous (135 million years ago to 70 or 65 million years ago, depending on which authority you follow). It took tens of millions of years, following the disaster which marks the Paleozoic/Mesozoic boundary, for life to become as varied once again as it had been before the disaster, and although land-based life had been less affected than life in the sea, the life forms that now diversified across the world were often very different from their Paleozoic predecessors, even though they often filled equivalent places in the hierarchy of life on Earth. By now, in particular, the amphibians had all but been wiped out, as much by the competition from reptiles as through any natural disaster, and were reduced to the few frogs and so on that we know today.

By the end of the Triassic, two separate groups of reptiles had become the two main branches of the dinosaur family. Throughout the whole of the Mesozoic, there were many occasions when waves of extinction swept across the Earth, and each time the largest dinosaurs died out. But each time, until 65 million years ago, they were replaced as selection pressures produced new large species from the surviving smaller varieties. But it is the big dinosaurs that spring to mind when the word is mentioned, and, indeed, the word means "terrible lizard" in Greek. A creature like *Tyrannosaurus*, six meters tall with a long tail and an impressive array of sharp teeth, was a terrible lizard indeed; but the name "dinosaur" also covers quiet, cowlike browsers and chicken-sized animals. It's rather as if all mammals were given the name "fierce cats" on the grounds that we are all related to lions and tigers; and a much better idea of life on Earth during the Mesozic is gained by looking at the diversity of animal life around today and imagining a reptilian equivalent for each land mammal. Dinosaurs were not just "fierce lizards," but spread everywhere that animal life spreads today—including the sea, where some species returned like dinosaur equivalents of modern whales and dolphins, which are also air-breathing descendants of former land-based animals.

MAMMALS UNDERFOOT

Alongside the dinosaurs—almost literally under their feet—there was a group of animals that had invented a new trick of reproduction, the logical next step (logical with hindsight, anyway) along the path the reptiles had followed when they developed an egg in which the early stages of development of a young animal could take place. Some reptiles evolved this development further, with the fertilized egg staying inside the mother's body instead of being covered in a hard shell and laid outside. All of the early development of the offspring could now take place within the mother's body, with the youngster only emerging into the outside world when it could literally stand on its own two (or four) feet. This method of reproduction has obvious advantages, and this particular evolutionary experiment produced the first mammals, our own direct ancestors, in the Triassic period. These small, hairy, warm-blooded animals were successful in a modest way right through the Mesozoic, as the fossil record shows, and spread across all the Mesozoic continents. But they could hardly be said to be competing with the dinosaurs, and were all rather like mice or shrews—no sign of a "fierce cat" as yet. Why didn't the obvious advantages of mammalian reproduction bring them more success during the Mesozoic? One reason is simply that the dinosaurs were established first. If a shrew had a genetic mutation which encouraged it to be fiercer, it might do very well competing against other shrews, but it wouldn't get very far trying to compete directly with a terrible lizard! It takes many generations for slight size advantages—taking just one example—to produce a big species from a small one, and this can only happen if there are no big predators around. If there are predators, then a big shrew is simply that much more noticeable and that much more tasty a mouthful. So all the selection pressures in the Mesozoic favored small, agile, fast-running shrew types, with good eyesight and hearing to detect predators.

But there may also be other reasons. In recent years there has been a considerable debate about whether dinosaurs were cold-blooded, like modern reptiles, or hot-blooded, like mammals.* The balance of evidence now is that at least some, probably many, and perhaps all dinosaurs were warm-blooded, since the structure of the bones of many fossilized dinosaur remains is characteristic of structures found in the bones of warm-blooded animals today. This would help to explain their success in spreading to every continent on Earth (except Antarctica) and adapting even to cold climates, where cold-blooded reptiles would be sluggish and inactive. It may also explain why the first mammals were not able to run rings around the dinosaurs, which could, if warm-blooded, have been every bit as active as mammals today. And it hints that the mammals may have been rather closely related to some dinosaurs, rather than just having common ancestry further back up the reptilian line of succession. Warm-bloodedness may have come first, before the splitting off of mammals from one of the two main dinosaur lines. There is also ample scope for speculation about whether all dinosaurs did lay eggs. Because of the resemblance to terrible lizards, the conclusion that they were egg layers was natural, and, indeed, fossil remains of dinosaur eggs have been found. But were *all* dinosaurs egg layers? Quite possibly, some had developed the mammalian trick of bringing forth their young alive, even if they didn't go the whole way with the "invention" of suckling and other mammalian characteristics.

The picture of animal life in the Mesozoic, then, may now be seen as very different from the popular image of giant lizards fighting with one another amid steaming swamps. The dinosaurs must, in fact, have been very similar to the diversity of warm-blooded animals around today, living the kinds of lives now lived by elephants, tigers, lions, cows, deer, and so on. The Mesozoic mammals, on the other hand, would have filled the "small animal" niches in the ecology, and would have been a significant element in the diet of at least some dinosaurs. The extinctions which repeatedly wiped out the largest dinosaurs, followed by the reappearance of new large species from the smaller dinosaur stock, continued in this picture until one extinction was so dramatic that

* See, for example, Adrian Desmond's *The Hot-Blooded Dinosaurs* (London: Blond & Briggs, 1977).

even the smaller dinosaurs were wiped out. But that still left the yet smaller animals, the mammals, and they spread out to occupy the large animal ecological niches, just as had happened many times before. Only this time the large animals too were now mammals. And in this picture the world today is very much like the world of the dinosaurs, but with some mammals having become "fierce cats" and the like, as replacements for the departed "terrible lizards."

But the dinosaurs have not gone entirely. One group of dinosaurs remains very much a feature of the world today. Birds are directly descended from flying dinosaur ancestors, and by any sensible definition they must be classed in the same overall family as those ancestors. The direct lineage of modern birds can be traced back to the dinosaur *Archeopteryx*, whose fossil remains have been found in mid-Jurassic rocks, and in the later Mesozoic there were many different types of bird diversifying and specializing to take advantage of the new possibilities opened up by flight. Some have given up flight, like the ostrich, and returned to a land-based existence. Some have gone further still and returned to the sea. The workings of evolution by natural selection are often convoluted and need not be logical. If you were setting out to evolve a creature adapted to swimming in the Antarctic Ocean, would you start off by designing a flying reptile? But however illogical it may seem, the only animals which had even a toehold on the Antarctic continent before man came along were the penguins—which, on reflection, means that the dinosaurs did eventually manage to spread even to Antarctica!

Such oddities emphasize how the evolutionary pattern has been shaped by the particular pattern of environmental changes on Earth since life appeared. If the pattern of continental drift, regroupings, and breakups had been different, and if the ebb and flow of ice ages had been different, there would still be a diversity of life on Earth today, but it would be quite different from the forms we see around us, even if fitted to the same ecological niches, just as a tiger differs from *Tyrannosaurus*. The particular pattern of environmental changes in the Mesozoic brought a first climatic transformation as arid desert regions were opened up to wet winds by the breakup of Pangaea II and the drift of continents

This reconstruction shows how Archeopteryx, the ancestor of modern birds, probably looked. Basically a flying lizard—a kind of dinosaur—Archeopteryx had teeth and a long, jointed tail. But, as fossil specimens 150 million years old show, it also had feathers. © *British Museum* (*Natural History*)

away from the icy polar region. Shallow seas came and went across regions such as the land which now makes up central Europe, while Gondwanaland and Laurasia become separate entities during the Triassic. Further rifting and drift of continents in the Jurassic saw the opening of what was to become the North Atlantic Ocean, and the split of South America from Africa, bringing spectacular upheavals along the western coast of North America and a shallow sea covering much of the continent while dinosaurs flourished in the warm, wet forests of the remaining land. In the Cretaceous, as the North and South Atlantic oceans joined and spread, climate was moderated by the proximity of the sea to most land masses; Gondwana had been completely split into the continents we know today, and Laurasia had all but ceased to exist, with only a narrow connection between the Eurasian and North American continents at the northern end of the growing Atlantic Ocean. And throughout the Mesozoic the climate seems to have been warm—certainly warmer than the cold which gripped much of the land mass of the world in the late Paleozoic. The ups and downs of different species, and the extinctions followed by reradiations, so important in the fossil record of the dinosaurs, can in many cases be linked directly with these physical changes in the terrestrial environment. But unlike the situation at the end of the Paleozoic, the changing terrestrial factors alone cannot easily account for the greatest of these extinctions, when at the end of the Cretaceous so many dinosaur species were wiped out that this time it was the mammals which radiated into every available animal niche. Something very dramatic happened a little over 65 million years ago, something which brought an end to the reign of the dinosaurs but which our ancestors were, fortuitously, well equipped to benefit from.

8 HUMAN ORIGINS

We owe our own origins directly to the events, whatever they were, which led to the demise of the dinosaurs. Mammals, our own ancestors, had been around for 100 million years or so before the abrupt end to the era of the dinosaurs, inconspicuously occupying small-animal niches in the ecology, scurrying about in the undergrowth. Small mammals may well have eaten dinosaur eggs, when they could get them, and medium-sized dinosaurs must surely have eaten small mammals, when they could catch them. So the two kinds of animal could be said to have been in conflict. However, there is no evidence at all that the mammals were the *cause* of the dinosaurs' demise. Mammals did not, as far as we know, suddenly become more skillful at finding dinosaur eggs and eat so many that the dinosaurs were wiped out. Nor can we make any claims that, at this time, our mammalian ancestors were any more intelligent than their dinosaur contemporaries. The traditional picture of the lumbering, stupid dinosaur with its huge body and tiny brain tells only part of the story. Many dinosaurs were stupid; many more were smaller, agile, and with respectably sized brains compared with their body weight, as well as probably being warm-blooded. While our ancestors were still shrewlike creatures hiding in the undergrowth, at least one group of dinosaurs had taken significant steps down the path which, with hindsight, we can recognize as leading to real intelligence.

These were the *Saurornithoides*, whose brains must have weighed about fifty grams, in bodies weighing about fifty kilograms, judging by the fossil record. This combination of brain and body weights is not so far from that found today in baboons, and *Saurornithoides* had a body rather suggestive of a small kangaroo with clawed feet and a long, flexible neck. The important point of this structure is that it freed the front limbs of the beast for holding things, and *Saurornithoides* had well-developed four-fingered hands. These features—upright walking, grasping fingers, a reasonable brain, and good eyesight—are ones which will become familiar in the story of human origins, and it is tantalizing to speculate on the future of intelligent life on Earth if catastrophe had not overwhelmed the dinosaurs some 65 million years ago. As Carl Sagan put it in *The Dragons of Eden*, if descendants of *Saurornithoides* had become the intelligent creatures dominating the Earth, among other things arithmetic with base 8 might have developed as "natural," since they had eight fingers to count on, while arithmetic with base 10 would seem quite exotic!

But the dinosaurs did die out, opening the way for mammals to radiate into the large-animal niches which the dinosaurs had occupied for so many millions of years. Disaster did strike, and almost certainly from outside the Earth, as a comparison with the previous great extinction* of species at the end of the Paleozoic soon shows.

DISASTER FROM ABOVE

The main difference between the earlier, Paleozoic extinctions and the ones that marked the end of the Mesozoic is that the species wiped out in the earlier disaster were mainly those that lived in the shallow seas, while at the end of the Mesozoic, al-

---

* As opposed to the *lesser* extinctions that took place *within* the Mesozoic.

though sea-based life was affected it was the large land animals that disappeared entirely. Sea-swimming dinosaurs were also indeed affected. But they were air breathers who had returned to the sea and swam on the surface, as unprotected from disaster from above as their land-dwelling cousins. The true fish, swimming deeper in the water, and bottom dwellers, seem scarcely to have noticed whatever it was that did strike down the dinosaurs. Sea creatures—especially the swimmers in the shallow seas around the continents—are exactly the species that ought to be affected by a pattern of the reconstitution of a supercontinent and ice ages, which we know occurred at the end of the Paleozoic. But there was no comparable coming together of continents at the end of the Mesozoic, and no great ice age. From the fossil record of the creatures of the shallow seas, there would be no reason to regard the time around 65 million years ago as a time of remarkable changes, indicating the end of the Mesozoic. Whatever happened overwhelmingly affected creatures on the surface of the Earth, especially large creatures, but had much less effect on anything below the surface of the sea—compelling evidence that the disaster struck the Earth from outside, from space.

There are three strong candidates for such a catastrophe. Two of them are variations on the same theme—the idea that, for one reason or another, the ozone layer which is so important for the protection of life at the Earth's surface was destroyed by some cataclysmic event, and that large animals were killed by a flood of ultraviolet radiation before it built up again.* The most dramatic version of this story, originally developed by the Soviet astronomer I. S. Shklovskii, lays the blame for the catastrophe on a relatively nearby supernova explosion, producing a flood of cosmic ray particles which plowed into the upper atmosphere, destroying the delicate balance of dynamic chemical equilibrium by which the ozone layer is maintained. The immediate burst of radiation would last only a few minutes, and the ozone layer could recover from that within a few decades. But David Clark, of the Royal Greenwich Observatory, has pointed out that an expanding shell of material blasted outward by the supernova would reach the

* Or the plants they depended on for food were killed by the ultraviolet radiation, and they died of starvation, which is just as effective.

Earth centuries after this initial burst of radiation, and would blanket the Solar System with energetic particles for further centuries, producing a much more long-term disturbance of the environment affecting life on Earth. Unfortunately for supporters of this dramatic idea, it seems that the chance of a supernova occurring close enough to Earth to do the job is very small. R. C. Whitten and his colleagues at the NASA Ames Research Center in California calculate that a supernova close enough to destroy 50 percent of the ozone layer happens only once every 2,000 million years or more. In other words, there is only a 1 in 100 chance that such a supernova occurred sometime in the past 20 million years, and about a 3 percent chance that one occurred sometime in the past 65 million years. Not impossible—but unlikely!

The alternative version of this idea sees the cause of the destruction of the ozone layer as a flicker in our Sun, a rare but not impossible outburst rather like the flares that we see today, but 100 times bigger. Our Sun is far from being a completely quiet star today, and shows several patterns of activity, the most relevant here being the familiar "sunspot cycle," roughly eleven years long, which sees a regular buildup of activity followed by a decline to a minimum activity state. At peak activity, the Sun experiences flaring outbursts which produce large amounts of particles, notably protons, which sleet across space in the solar wind, some of them arriving at the atmosphere of the Earth. G. C. Reid and a team working at Boulder, Colorado, have calculated the effects of solar protons on the ozone layer and thereby on life on Earth. Because one effect of the solar protons is to encourage production of nitric oxide in the stratosphere, and nitric oxide itself combines with ozone to produce nitrogen dioxide and "ordinary" diatomic oxygen molecules, the protons are very good at reducing the concentration of ozone in the stratosphere. At least they would be good at destroying ozone if they could penetrate easily into the ozone layer. Fortunately for us, there is another layer of shielding around the Earth which protects us from the worst effects—the Earth's magnetic field, which deflects the electrically charged particles of the solar wind.

But the magnetic field is not always there to protect the stratosphere from charged particles. Remember that the geomagnetic field reverses from time to time, and that during reversals it is, at best, very weak for hundreds or thousands of years. During that time even an ordinary solar flare of the kind familiar today could reduce the ozone content of the stratosphere so much that 15 percent more ultraviolet would penetrate to the ground. A flare 100 times stronger than usual—quite likely sometime in the span of a few thousand years—would double or triple the amount of ultraviolet radiation getting through to the surface of the Earth. This process may well be important as an influence on life on Earth, and may well have played a part in at least some faunal extinctions, since the fossil record does show some tendency for extinctions to occur at times of geomagnetic reversals. But it is far from clear that it was the dominent influence at the end of the Cretaceous, when the dinosaurs were wiped out.* At present, the best candidate for the outside influence which led to their downfall seems to be literally an impact from outside—the collision of a large meteorite with our planet.

The Earth has been bombarded by meteorites since it formed—indeed, it formed out of the cloud of material around the Sun as a result of collisions between lumps of material in which the biggest lump grew to be a planet. Over the past 4,000 million years or so, the amount of nonplanetary material in the Solar System has declined, as more and more lumps have been swept up by the planets. The battered surfaces of the Moon, Mercury, Mars, and the moons of Jupiter, familiar now from photographs returned by space probes, show what a pounding every planet or moon has received in the process, and the craters are less obvious on Earth only because they have been softened by erosion and destroyed by tectonic activity. (On Venus, the craters may be there, or may have been eroded, but in any case the surface is hidden from us by thick layers of cloud.) Even during the present geological era, the Cainozoic, over a span of only 65 to 70 million years, half of all the ocean floor—one-third of the surface of the whole planet—has

---

* The circumstantial evidence, however, is not entirely absent. Could it be relevant that, according to most interpretations of the evidence, our small mammal ancestors occupied the ecological niches typical of nocturnal creatures? Were the dinosaurs, active by day, thereby exposed more directly to changes in the nature of the solar radiation reaching the ground? I am not convinced, especially since if some dinosaurs were warm-blooded they could also have been nocturnal. But it is an interesting argument!

been renewed by spreading from ocean ridges while old ocean crust has been recycled through deep trenches. Only the land surface provides any record of events more than 150 million years old, and the land surface has been considerably reworked. Even so, some dramatic features can still be identified, and the youth of a few of them indicates that although the extraterrestrial bombardment has diminished since the Earth was young, it is still not entirely over.

Barringer Crater in Arizona is the classic example. Three-quarters of a mile (3937 feet) across and 600 feet deep, it was produced by a meteoritic impact which can be dated, by standard geological techniques, at only 25,000 years ago. Vastly greater features, such as West Clearwater Lake in Quebec (thirteen miles across) and the Vredefort Ring in South Africa (thirty-five miles across) show the characteristic circular shape of meteorite impact features, and are almost certainly huge craters produced by impacts hundreds or thousands of millions of years ago. Such impacts must have had a pronounced effect on the environment, at least locally, and the bigger ones would have had a bigger and more widespread effect. At the end of 1979 two scientific teams independently came up with the suggestion that a particularly dramatic meteorite impact might have changed the terrestrial environment sufficiently to account for the extinction of the dinosaurs; unknown to either group, however, the idea had originally been proposed by J. I. Enever as long ago as 1966. Enever's "mistake" seems to have been to publish his calculations in a nonfiction article in the magazine *Analog*, better known for its science fiction stories. However, in his honor it seems appropriate to dub the idea "Enever's hypothesis."

Enever started with some very simple calculations of the energy involved in the impacts which have produced features such as the Vredefort Ring. The equation involved is one of the simplest in physics—a body of mass m moving at velocity v has a kinetic energy of $\frac{1}{2}mv^2$, and if the body is brought to a halt by an impact (or any other means) that energy has to go somewhere. Usually it is converted into heat, which is why the brakes and tires of a car get hot when you stop the vehicle. A fairly ordinary meteoritic body might be moving through the Solar System at a speed of

Barringer Crater, Arizona—the classic example of the scar produced by the impact of a meteorite with the Earth. Could such an event, on an even larger scale, have caused environmental changes that led to the death of the dinosaurs and the end of the Mesozoic Era, some 65–70 million years ago?

about 50 km per second, relative to the Earth, and might have a mass of thousands of tons. If such an object hit our planet, the kinetic energy converted into heat would be equivalent to the explosion of 100,000 megatons, or more, of TNT—bigger than any nuclear device yet tested by man. Even this, however, would not be enough to explain features as large as the Vredefort Ring. For that the impact required must have produced 10 *million* megatons' equivalent and, using the kinetic energy equation, this implies a mass of 32 thousand million tons—comparable to an asteroid such as Hermes, and by no means impossible in the Solar System even today—coming in at a speed of about 50 km per second. Enever's special contribution, though, was to speculate on what might have happened if a comparable meteoritic strike had happened not on land, as in the case of the Vredefort Ring, but at sea.

It might seem that an oceanic strike would be less spectacular than one on land, since it would be "damped" by the water. But that is entirely wrong! Quite apart from the tidal waves produced, which would ravage coastal regions, the almost unimaginable amount of energy released as heat in the impact would not only vaporize the water of the sea at the point of impact, but would punch a hole scores of kilometers wide right through the thin crust of the ocean floor. Seawater pouring into this pit would eventually cool the exposed molten rock and restore normality—but with 16,000 cubic kilometers of water, on Enever's calculations, vaporized to steam along the way. This would be enough to cover the entire surface of the Earth with two centimeters of rainfall— or, rather, to provide a temporary cloud cover reflecting away so much solar radiation that the whole Earth would be covered with a temporary blanket of snow! At the same time, the rising fireball of superheated air from the impact would punch its way upward into the stratosphere, disrupting the ozone layer as efficiently as any solar flare. The combined effects would, clearly, be very serious for land-based life, but considerably less serious for fish. And although the idea is speculative, it is certain that impacts of this kind must have occurred in the oceans during geological history—after all, two-thirds of the Earth's surface is sea, and the Vredefort Ring and other features show the evidence of similar events on land.

into the calculation, with one group from the Scottish Royal Observatory pointing out that big meteorites are more likely to strike the Earth when the Solar System is passing through a spiral arm of the Galaxy and being subjected to encounters with all kinds of cosmic junk. Since we have passed through a spiral arm quite recently, by astronomical standards, this looks interesting. (I should also mention that more supernovae occur near spiral arms, so this is just as interesting for the alternative death-of-the-dinosaur hypothesis!) The other group, an American team headed by Luis Alvarez, came up with something even more concrete, and announced their results, producing a wave of publicity, at the

1979 meeting of the American Association for the Advancement of Science in San Francisco.

Like others before them, the Alvarez team cited the example of the Vredefort Ring, and calculated that such an event should throw a blanket of fine dust high into the stratosphere. Eventually this dust will settle as a thin layer across the surface of the Earth, but first it could precipitate a crisis for life on Earth by blocking off enough of the Sun's heat and light to kill many plants and the animals which depend on those plants for food. This version of the idea doesn't even require the metorite to strike at sea, and it is backed by hard evidence in the form of isotopes of iridium found in rock layers 65 million years old.

Strictly speaking, the discovery of iridium came first—the Alvarez team was looking for traces of meteoritic dust, but hadn't initially thought of the link with events 65 million years ago. Iridium is a good indication of the presence of extraterrestrial material, since there is very little in the Earth's crust. So the discovery of a thin iridium layer in strata exactly coinciding with the faunal extinctions of 65 million years ago immediately made the team think of a link between extraterrestrial events and the extinctions. So far, these results have not been published in detail, and the immediate test of the idea will come with a search for the iridium traces, so far identified only in strata from Italy, in cores drilled from both land and seafloor around the world.*

SMALL SURVIVORS: A NEW EPOCH

Whatever the exact cause, though, the fact is that the dinosaurs were wiped out by a catastrophe which left only animals weighing less than about 40 kg as survivors.† Our tiny ancestors were well within this weight limit, and several lines of mammals survived the death of the dinosaurs and took advantage of the opportunities provided by their departure to spread and diversify. Even the term

* The Alvarez team's report appeared in *Science* (vol. 208, p. 1095) in June 1980. By the end of August several other groups had reported confirmation of traces of extraterrestrial material in rock samples 65 million years old from Denmark, New Zealand, and Spain. There is now very little doubt about the giant meteor theory as an explanation of the Cretaceous/Tertiary boundary event.
† *Saurornithoides* very nearly did survive, on this evidence!

The primate line is descended from small, agile mammals like this tree shrew. Even *Aegyptopithecus*, one of the first apes to evolve from the monkey line some 30 million years ago, was no bigger than a large modern house cat or rabbit. © *British Museum (Natural History)*

"mammal" is already too general when tracing the story of human origins forward from 65 or 70 million years ago; the narrower definition "primate," the biological "order" which includes man, seems more appropriate for some mammals which had already appeared while the dinosaurs still dominated the Earth, and from 65 million years ago we are talking about primate evolution rather than even the evolution of mammals as a whole.*

Once again the focus of attention in our search for human origins has been refined, and once again the changing focus is reflected in the standard geological timescale. Although the era in which we are living, the Cainozoic, began only some 70 million years ago, we have much more detailed evidence from the fossil record about Cainozoic events than about the preceding 160 mil-

* At least according to some authorities. Others would argue that the particular insectivorous, shrewlike mammals scurrying around in the late Cretaceous weren't yet quite primates, but James Valentine, professor of geological sciences at the University of California at Santa Barbara, is one of the experts who would place primate origins before the death of the dinosaurs. See, for example, "The Evolution of Multicellular Plants and Animals," in *Scientific American* 239, 3 (September 1978):105.

lion years or more of the Mesozoic, just as the information we have from the Mesozoic is more detailed than that from the even longer era, the Paleozoic, which preceded it. This does not mean that recent events have been more dramatic or important than the earlier history of the Earth; it's just that we know more about them. By the Cainozoic, indeed, we know so much that the division even into geological periods is inadequate as a calendar against which to note the changes in our planet and our ancestors, although the Cainozoic is subdivided into the Tertiary, which ended about 3 million years ago, and the Quaternary period, in which we are living. The geological *epochs*, yet finer subdivisions of geological time, are much more important for the rest of the story.

It also makes sense, for the first time, to set the epochs in context by counting backward from the present day. The Holocene, or Recent, epoch began only 10,000 years ago, at the end of the most recent great Ice Age. This is a rather unnatural division in one way, since we are almost certainly living in a minor interglacial within a longer span of ice age conditions, so that in terms of the physical conditions on Earth the start of the Holocene marks no special event. However, it is linked with a biological event of crucial importance to us, the beginnings of human civilization with the invention of agriculture. So it is certainly a sensible subdivision in human terms. The previous epoch, the Pleistocene, covers the rest of the 3 million years of the Quaternary, and is preceded in turn by the Pliocene, which began 7 million years ago; the Miocene, which began 18 million years ago; the Oligocene, 15 million years long and beginning 40 million years ago; the 20 million-year-long Eocene; and the Paleocene, 10 million years long, which was the first Cainozoic epoch and followed the end of the Mesozoic, some 70 million years ago. The numbers can be put in perspective by the realization that the entire Cainozoic, during which half of the oceanic crust of the world has been recycled, corresponds to a time span just one-quarter of that covered by the story of the reptiles, and less than 2 percent of the entire history of the Earth.

So close to the present day it is possible to put all of the significant developments of life into a paragraph or two. During the

Paleocene and Eocene, flowering plants became dominant and deciduous trees first became prominent, while the ancestors of the whales left the land and went back to the sea from which their fishy ancestors had originated. On land, as well as the early primates, the immediate ancestors of the elephant, horse, crocodile, and tortoise, among others, had appeared. In the Oligocene, taking us up to 26 million years ago (26 Myr BP, for "Before the Present"), the European Alps were building up as Africa began to squeeze up toward Europe, and ape fossils were being laid down in some parts of the world.

Many mammals, including the apes, spread out during the Miocene, and species diversified. In the Pliocene, however, conditions changed and many of those species were thinned out in a series of extinctions culminating in those associated with the recent wave of ice ages. There used to be many more mammal species around than there are today; but the survivors are, by definition, the "fittest" in Darwinian terms, and the primate line that led to man seems to have been particularly well suited to surviving the ice age rigors.

But if such a brief discussion of the events of the past 70 million years is appropriate in the context of the 4,500 million-year history of the Earth, it is hardly appropriate in terms of human history and origins! So let's look at the story in a little more detail.

It isn't always possible to pick out from the fossil record each step down the evolutionary path that led to man, but it is possible to pick out the features of those shrewlike insectivores which made them front-runners in the bid to fill some of the ecological niches left vacant by the departure of the larger dinosaurs. Many of the things which made our species so successful in the past few million years were a result of our immediate ancestors moving out of the trees and adapting the skills developed in the forest to a life on the plains. But at the time we are now considering, 70 million years ago, our more distant ancestors were making the opposite step! The small mammals scurrying about on the floor of the forest, living off a diet of seeds and insects, moved into the trees as the opportunity opened up, and started looking for seeds, insects, and anything else they could eat up in the branches. Life in the branches required several significant adaptations that remain a feature of primates today, including ourselves.*

LIFE IN THE BRANCHES

The first necessity was hands capable of holding onto the branches and of picking up small food items such as insects—hands with nails, rather than claws, better adapted both to grasping and to delicate tasks. The second necessity was good eyesight, in three dimensions. This stereoscopic vision is provided by eyes at the front of the head, working together. And the third required characteristic was the ability to sit or stand upright, freeing the front limbs for grasping and manipulating food. Only primates have all three of these characteristics, although other mammals may have one of them, and all predators, for example, have stereoscopic vision.

In the branches the fittest for survival were both acrobatic, with a fine sense of balance and superb reactions, and capable of very detailed "work" coordinating the eyes and the hands, through the brain, to pick up the insects that were still a staple diet of our ancestors. Also, the movement into trees and the demise of many former predators allowed some of the descendants of the nocturnal first primates to change to a life of daytime activity—and in the brighter light of day, the evolution of color vision was a kind of bonus; it wasn't essential to life in the trees, but something which made life that much easier, and opened up completely new possibilities for identifying other creatures, including individuals of the same species, and eventually for communication. Since color vision helps in identifying at least some kinds of food—such as fruit—the advantage of color vision is so enormous that any genes which tailored the eyes to see in color were bound to sweep through the gene pool rapidly. And this ability, once again,

---

* Strictly, "members of the species which possessed genes that produced those adaptations were able to move into the trees, getting progressively better at the job of living in the branches as natural selection encouraged the development of the advantageous features." The insectivores didn't take a quick look at a tree, say, "Hm, I'd need good eyesight to live up there," and promptly evolve good eyesight. Rather, shrews with good eyesight got more food and reproduced effectively, while those with bad eyesight didn't get much chance to pass on their genes.

requires an efficient means of processing the available information—a good brain.

But all of this took time. Indeed, it took 30 million years of evolution to equip the descendants of the original tree shrews with all the paraphernalia of modern primates outlined above, while the primate order diversified into a variety of tree dwellers, most of them nocturnal and most still living off a diet of insects. These prosimians, as they are called, still survive in some parts of the world, notably Madagascar, where they have been cut off from mainland Africa for scores of millions of years and have never had to compete with the more advanced primates that evolved later. The more advanced primates, the anthropoids, evolved from one of the prosimian lines, and the first of them, the earliest monkeys, appeared about 40 million years ago. Their existence then has been revealed by fossil traces found in strata about 100 kilometers south of Cairo in Egypt. The arrival of those first monkeys was bad news for the prosimians. Larger and more intelligent, the monkeys quickly displaced the more primitive primates from most of the daytime tree-dwelling niches, while they also made the most of the diet of leaves and fruit available in the trees, leaving the insects for an occasional snack, or for the nocturnal prosimians.

Such an evolutionary development was inevitable, given enough time for the constant shuffling of genes to produce a creature even better adapted to life in the trees. But around 40 Myr BP the climate of the Earth was already changing, after a long period of "tropical" conditions over most of the land. As the continents drifted toward their present-day positions, the presence of land at high latitudes and the cutting off of warm water from the polar regions began to be important, and the worldwide climatic changes that began to be significant about this time may have played a part in the hastening evolution of the primates, by making conditions harsher and thus making it more difficult for less successful variations on the theme to survive. This wasn't the only influence of continental drift on the course of primate evolution. By the time monkeys appear in the fossil record, Africa and South America were well separated, and it seems unlikely that early monkeys from Africa could have made the voyage, even across the then-narrow "ocean," from Africa to South America. Yet there are

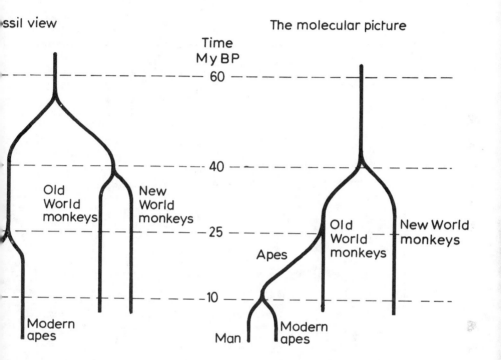

ssil view   The molecular picture

Time
My BP

60

40

Old   New
World   World
monkeys   monkeys

                                    Old   New World
                                    World   monkeys
                              25   monkeys

                        Apes

10

Modern   Modern
apes   Man   apes

The most striking puzzle of evolutionary studies today is the disagreement between the timescales of human evolution calculated from the traditional fossil evidence or from the new molecular evidence. Did the monkey/ape split come
before Old and New World monkeys diverged, with man and ape lines separating
25 million years or more ago? Or did apes split only relatively recently from the
Old World monkey line, with the man/ape split within the past 10 million years?
There is actually more evidence to back up the molecular picture—the fossil
remains of our distant ancestors are few and far between. But the traditional picture is well established as the accepted view, and at present efforts are being made
to find a compromise timescale using the best bits of both approaches, rather than
to throw out the fossil evidence, scarce though it is, in favor of the molecular
clock.

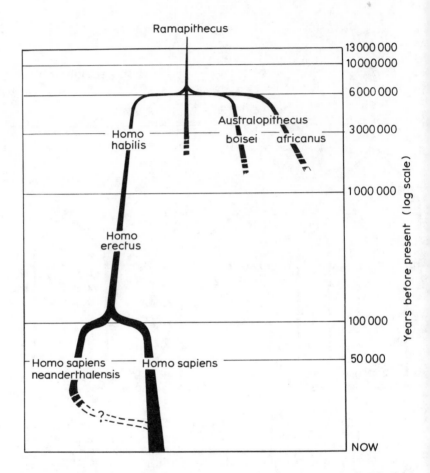

Accepting the evidence that *Ramapithecus* was the direct ancestor of man, the probable evolutionary path of the hominids can be traced back for more than 10 million years. In earlier versions of the story, our australopithecine cousins were placed at intermediate stages on the direct route from *Ramapithecus* to *Homo sapiens*; today, the best evidence is that they lived alongside our direct ancestors, a separate branch from the early ramapithecine stock. This is not the final picture—further studies of the molecular clock, in particular, are likely to lead to at least some changes in the near future. But it is the best "family tree" for man that it is possible to draw at present.

monkeys in South America, and at first sight they look just like their cousins in Africa. However, there are important differences in their teeth and other features, including the arrangement of nostrils, and unlike the Old World monkeys of Africa and Asia the New World monkeys have the ability to use their tails almost as fifth limbs for holding onto branches—Old World monkeys use their tails only for balance. Almost certainly the two kinds of monkey show a striking example of parallel evolution from an early common ancestor, something intermediate between tree shrew and monkey which lived in both South America and Africa before they split. Each line of evolution then selected the best features for the job of living in the trees and developed them from the pool of genetic material available, coming up with the "answer" monkey. But in the New World the line stops with monkeys—the evolutionary pressures that produced further adaptations in the Old World, leading to the first apes and eventually to humans, simply weren't present in South America, where the first apes on the scene were human beings who arrived only a few tens of thousands of years ago from the Old World.

ARRIVAL OF THE APES

So what were the factors that encouraged further evolution of the primate stock in the Old World, producing the first apes only some 10 million years after the first monkeys appeared? The best explanation is that in South America thick tropical forest continued to dominate the scene, and the monkey way of life was well adapted to such forest conditions. South America was shifting mainly westward in the dance of the continents, always near to the tropical region of the Earth. Africa, on the other hand, was shifting northward, and the great forests were in decline. A slightly drier climate and the presence of areas of more open woodland were enough to provide a slightly different kind of home for life, and as ever, life adapted to fit the new ecological niche. The result was the appearance of the apes on Earth by about 30 million years ago. And from this point on, by any definition, we are talking about the origins of man.

The oldest fossil remnant identified as belonging to an ape

The chimpanzee, our closest relative, is almost uncannily human in many respects. Yet it has still come as a surprise, not yet satisfactorily explained by evolutionary theory, to discover that the chimp and ourselves share 99 percent of our genetic material.

comes from the same region of Egypt as the oldest monkey fossils, the Fayum Depression, south of Cairo. This is a desert region today, but 30 or 40 million years ago it was covered by lush forest. Details of how paleontologists identify particular fossils as those of an ape rather than a monkey, or of a particular ape rather than one of its cousins, make fascinating reading, as do the stories of the arguments between paleontologists over such identifications. But this is one of those occasions where the details have to be skated over for brevity, and we have to accept that the features of the 28-million-year-old skull, almost complete except for the jaw, are unambiguously apelike to the experts. So this is the oldest direct ancestor which we share with other living apes such as the chimpanzee and orangutan. The name given to this ancient ape is *Aegyptopithecus*, from its location in modern Egypt and the Greek *pithecus*, meaning "ape." Just as monkeys are bigger than prosimians, so apes are bigger than monkeys, in general, and these differences are related to their different ways of life. The prosimians scurry along the branches, while monkeys, with their powerful hind legs, jump from bough to bough and tree to tree. The apes are too big for such leaping, and swing from branch to branch using all four limbs, effectively using four hands for the job. Swinging underneath a branch, an ape has three hands to hold on with and one free hand for picking fruit.

Of course, with such closely related species the distinctions are blurred. Some large modern monkeys have taken up the way of life typical of the early apes, while man is not the only primate to have investigated life on the ground. But the broad differences indicate how the species diversified into slightly different roles.

With the arrival of the apes on the evolutionary scene, we have come to yet another subdivision of species, the hominoids, which are represented only by man and the other apes. The distinctions between species at this level are so subtle, and the fossil record so incomplete,* that it is hard to be absolutely certain of the details of how the descendants of *Aegyptopithecus* became us, but it seems

---

* One reason for the paucity of hominoid fossils is that our ancestors lived in wooded regions rich in life, where animal remains didn't have a chance to lie around and get buried before being eaten by something or rotting. This problem, indeed, applies to the whole primate story, which has to be built up from fragments of bone, teeth, and the occasional

that shortly after the first ape appeared the hominoid line diverged into two main branches, one represented by fossils of the genus *Pliopithecus*, the ancestor of the modern gibbons. The other line, *Dryopithecus* (meaning "tree ape"), seems to have been widespread in the Miocene, and was certainly established by about 20 million years ago (20 Myr BP), 8 million years after the appearance of *Aegyptopithecus* in the fossil record. *Dryopithecus* fossils have been found from Africa, Asia, and Europe, with the most complete remains coming from the Lake Victoria region of East Africa, where they were excavated by Louis and Mary Leakey. The true *Dryopithecus* genus is the direct ancestor of modern species such as the orangutan and gorilla, but probably not of man. It seems that we share a common ancestor—not yet found—best called a "dryopithecine," which gave rise to *Dryopithecus* and two other lines. One of these, *Gigantopithecus*, was the ancestor of a family of large apes that used to inhabit Asia and is now extinct; the other, *Ramapithecus*, seems to have been our own ancestor. Frustratingly, though, there is a gap in the fossil record from the time of *Ramapithecus* (10 Myr BP) up to about 5 million years ago. This may seem a small gap, compared with the broad sweep of time covered since the origin of the Earth, or even of life on land. But in that crucial gap the *Ramapithecus* line, learned, among other things, to walk upright! Small wonder that paleontologists are so eager to find ancestral human (hominid, as opposed to the ape [hominoid] lines) remains from the period between 4 and 8 million years ago.

BROTHERS UNDER THE SKIN

Our arrival at this still crucial gap in the fossil record is an appropriate place to pause and take stock of how certain we are that the story just outlined is correct. Certainly all the pieces hang together, suggesting that our own ancestors had diverged from the other ape lines by about 15 million years ago, with the arrival of *Ramapithecus*. But there is no proof of unbroken descent from

---

skull. Our ancestors, unfortunately, were not in the habit of lying down to die in the silty waters of river banks—although some all-important discoveries have been made from just such locations.

Ramapithecus to us, and there is intriguing new evidence that the split between human and ape stock may have happened much more recently than 15 million years ago. It all boils down to how closely we are related to different apes—and molecular biology is now a sufficiently precise science to be able to compare the proteins in the DNA of human cells and ape cells and to determine just how alike we are.

The results of this work, pioneered by American biologists Vincent Sarich and Allan Wilson,* are surprising in the light of the conventional interpretation of the fossil record. The new evidence clearly shows that our closest relatives are the gorillas and the chimpanzees, with orangutans more distantly related and gibbons more distantly still. So far there is no great conflict with the fossil evidence. But in genetic terms, the differences between a man and a chimpanzee, or between man and gorilla, are *much* less than the differences between an orangutan and a chimpanzee. This tells us one very important thing—since chimpanzee and gorilla are exclusively African animals, there can be very little doubt that human origins lie in Africa. We simply are not closely enough related to the Asian apes to have evolved there. But the first surprise this work throws up is its completely different indication of the timescale of human evolution.

Clearly, if the chimpanzee, gorilla, and man had a common ancestor, then the differences between the genetic material of the three genera today have been built up by mutations since the time of the common ancestor—crossovers which changed the genes slightly and so on. The longer the time since the common ancestor, the more genetic differences there should be between the diverging species, since there has been more opportunity for mutations to occur. By looking at the protein divergence—the differences between genetic material—in different species of mammal, the molecular biologists have developed what seems to be a reliable guide to the rate at which divergence occurs, so that they ought to be able to use these studies as a "clock" to determine just when the human stock did split off from the line leading to other modern apes. But simply applying the "clock" worked out

* And superbly explained by Colin Patterson in *Evolution* (Boston: Routledge & Kegan Paul, 1978).

from other mammals suggests that even the gibbon separated from our evolutionary line only 10 million years ago, the orangutan 7 million years ago, and the chimpanzee and gorilla as recently as 4 million years ago. But by 4 million years ago, the fossil record shows, there were very manlike creatures around. There is a clear conflict between the fossil clock and the molecular clock, and something has to give.

Perhaps the best explanation is that men and apes have evolved more slowly than the other mammals, against which the clock was calibrated. This could be because we have a long generation time, a broader gap between parents, children, and grandchildren than most mammals—after all, the more generations there are in a given time, the more chances for mutations to occur. The molecular clock, it seems, measures time not in years but in generations. Another possibility is that *Ramapithecus* was not our direct ancestor after all and that the molecular clock is giving us a much more accurate version of the timescale than the conventional one based on the fossil record. At present, the dilemma remains, and the best we can say is that the fossil record tells us that it is highly unlikely that the human line separated from chimpanzee and gorilla lines less than 5 million years ago, while the molecular clock indicates that it is highly unlikely, even allowing for these special factors, that the split occurred more than 10 million years ago. The leeway is just sufficient to keep most of the authorities reasonably happy, and we can say that sometime in the past 10 million years our ancestors, possibly descended from *Ramapithecus*, took the next step along the path, out of the forest which had served the primates so well since the demise of the dinosaurs some 60 million years earlier still.

But what happened to make us human was, by any standards, a very peculiar evolutionary development. The peculiarity is only emphasized by the new molecular studies, since the second big surprise which they spring is the closeness of our relationship with gorillas and chimpanzees. In fact, on the basis of comparisons of the protein chains, man and chimpanzee share 99 percent of their genetic material.* The genes which make us human rather than

* According to studies by Marie-Claire King and Allan Wilson (see Bibliography).

chimpanzee are only 1 percent of our total stock! Whether or not the molecular technique is telling us anything useful about the timescale of evolution, this is one of the most astonishing discoveries in the whole story of evolution, and it is perhaps just as well for their own peace of mind that the more virulent critics of Darwin at the time of the publication of his *Origin of Species* were spared this knowledge. It means, in simple terms, that humans and chimps differ less than some species that are placed in the same genus as each other. A fox and a dog, for example, are shown by the new technique to be *less* closely related than man and chimp in spite of their similar appearance. We classify animals today into categories by their ability to mate with one another and produce offspring (of course), and by their physical appearance. Certainly man and chimp are much more different physically than a 1 percent difference in genetic material would suggest. This is not just a subjective view; measurements of the proportions of human and chimp bodies (length of arms compared with length of body, width of head, and so on) show real differences in the two. But there is much less difference, in these terms, between a baby chimpanzee and an adult human. It seems that the small genetic difference between man and chimp is so important because the genes which have changed are the ones which control the growth and development of the body; man is a kind of "immature" or "infant" ape, which never develops fully compared with its cousins—a phenomenon known as neoteny.

THE INFANT APE

Biologists had recognized this already, and realized that a great many of the features which make us human, and which made our ancestors so successful when they came out of the forest, can be described in neotenous terms. The phenomenon is known in other species, and it can have a very powerful effect in evolution because, precisely as has happened with man, mutations in a few genes can have a dramatic impact on the overall structure of the body they inhabit. The crucial question then is why the neotenous ape—man—should have been so successful, and nowhere has this question been more clearly answered than by Desmond Morris in his book *The Naked Ape*, published several years before the discovery of the close genetic similarity between man and chimpanzee, the realization that we are almost literally "brothers (or sisters!) under the skin."

With climatic changes that brought a decline in the tropical forests, a creature such as *Ramapithecus* living 10 to 12 million years ago was in trouble. The most successful offspring—the fittest in evolutionary terms—would be the ones that coped best with the changing conditions, and that meant moving out of the trees and back on to the ground, into the spreading plains of East Africa. But the original primate skills from an earlier phase of land-based life had by now been lost—the most obvious and important example, perhaps, is the sense of smell, which must have been of prime importance to our shrewlike ancestors snuffling about among the dinosaurs, but had been largely replaced by improved eyesight in the tree-dwellers. Given time, evolution could undoubtedly have produced a line of *Rampithecus* descendants with an improved sense of smell, running about on all fours to keep their noses close to the ground. But the changes produced by neotenous development are much quicker than other evolutionary changes, hardly requiring any genetic mutation at all. In any species there is a spread in the rates at which individuals mature, with some taking longer than others. So any selection pressure which favors slow physical development can work immediately on the population, with slow developers surviving better and reproducing more effectively. In the new population of their descendants, there is a new spread of maturing rates, now centered on a slower "normal" rate, and the whole process repeats itself immediately in every generation. Only a few genes, the ones which control the development and rate of growth of the different parts of the body, are involved, and if there is any advantage in slower development rates, then this shows up immediately and effectively— rather in the way that the gene which controls the number of eggs laid in a clutch is selected very effectively in a bird population.

Of course, there is a need for the slow-maturing infant to be protected by its parents or other adults, but that is a feature of all apes and must have been part of the pattern of behavior of *Ramapithecus* too. And then it is necessary that neoteny should be advantageous; the best way to see why it was is to imagine the design problems involved in adapting a tree-dwelling ape with good eyesight, well-developed hands, and a large brain for life on the plains.

The tree-dweller's greatest asset in those circumstances would be versatility. A species which had evolved into complete dependence on trees would die out with the forests, but one which could literally *learn* new tricks, instead of just waiting to evolve, would have a better immediate chance of survival. So the brain must have been an asset right from the start of this forced migration onto the plains. Most animals complete the bulk of their brain development before birth, and even our nearest relations, the chimps, complete brain growth by the end of the first year of life. In humans, though, the brain is only a quarter of its final size at birth, and growth continues for nearly twenty-five years. This slow development is a feature of neoteny; it means, in practical terms, that the brain can grow to a size that would be impossible before birth, since if all this development took place in the womb then the baby would simply have too big a head to be born without killing its mother. One result of this continuing growth of the brain is that infants and even young adults are out in the world and learning about it while the brain is still growing—which probably explains the adaptability of modern man. The striking contrast is with offspring of animals such as deer, which can run about with the herd almost as soon as they are born, and have brains "preprogrammed" to cope with herd life, but can never learn anything else, being set in their ways by the preprogramming.

What else did our ancestors need? It must certainly have been an advantage to be able to stand upright and use their improved eyesight to scan their surroundings for danger, and this requires the head to be positioned in an unusual way compared with most mammals. If you try crawling about on all fours, you will soon see that the natural position of the head is looking straight down at the floor. Four-footed mammals have adapted to their way of life by evolving a head which is tilted at right angles on the neck, so that they look forward; but if such an animal (a dog, perhaps) is trained to stand on its hind legs, then it is looking straight up at the sky. Mammal fetuses, mimicking the history of evolution, go through a stage where the head is pointing at right angles to the trunk, like your head and mine. And the way we hold our heads today is a reversion to that fetal stage, another example of neoteny.

Many other features can be picked out and analyzed in this way, but it is surely sufficient to realize that by one relatively simple evolutionary step, neoteny, our ancestors gained the powerful advantages of better brains, a body better suited to upright posture and able to run about with eyes alert for danger and hands free to carry food or weapons, and a longer childhood during which the developing brain could be taught an appropriate way of life by the parents. Two side effects, one trivial and the other perhaps of overriding, if accidental, significance, should be mentioned.

The first is that we are relatively hairless—the "naked ape," in Desmond Morris's memorable term. A fetal chimpanzee, which looks so much like a human, even down to the round shape of the head, and the proportions of limbs, hands, and feet, also has very little hair except on the head and chin. Our "nakedness" is simply a side effect of neoteny, and probably doesn't have any particular advantage in terms of natural selection, being just one of the things that came in the parcel of neotenous development, with no strong disadvantage which would have caused genes for nakedness to be selected against. The entertaining debate about why a naked skin should have evolved may be a complete red herring, no more relevant than asking why we should have five-fingered hands instead of four.

Secondly, though, neoteny may have led directly to the feature of human behavior which caused our species to spread across the world. Infant apes are noted for their curiosity and playfulness, but in all cases except man this phase of inventive exploration dies out fairly quickly. Whatever the direct cause of such behavior—probably linked with the time it takes for the brain to grow—humans retain this streak of curiosity well into adult life. This inclination to find out what lies just over the next hill was all the drive that our ancestors needed to conquer the world. Both the fossil evidence and the comparison of our genetic material with that of modern apes shows unequivocally that our immediate ancestors came from one part of the world, East Africa. But they didn't *set out* to "conquer the world" or anything like that. Armed just with curiosity, a few members in each generation must have wandered off over the hills and set up home a few miles away. Some of their descendants, in turn, must have wandered over the next range of hills before settling down to raise a family. And at a rate of only ten miles per generation, the journey from Nairobi to Peking would take less than 15,000 years—a mere eyeblink in evolutionary terms, with people scattered all along the route, and spreading out in other directions, to develop in due course the evolutionarily trivial local characteristics which distinguish a black African from a yellow Chinese, an Arab from an Eskimo. The human world today is as it is because of the unique human combination of intelligence and inventive curiosity; we are what we are, and the world is the way it is today, because man is the ape who never grew up.*

OUT OF THE TREES

All that, however, was still far in the future when the *Ramapithecus* line—assuming that *Ramapithecus* really was our own ancestor—came out of the woods. It's just as well, perhaps, that we can understand so clearly the nature of the evolutionary process which turned a tree-ape into man, since there are no fossil hominid remains yet known for the period from 10 million years ago to 5 million years ago, and then nothing except one jaw fragment, which the experts are still arguing over, until 3 million years ago.† In terms of the fossil record, we see the *Ramapithecus*

---

* Carl Sagan, among his more serious comments on the evolution of human intelligence, throws out the delightful idea that some kind of ancestral memory of days gone by may account for many human fears and myths. The dragons, of course, are the dinosaurs who were for so long our rivals, and the large reptiles who, he suggests, men may have had direct conflict with much more recently than 65 Myr ago. Not long before our own line became the single human species on Earth, there were races of both small and large "men," and Sagan speculates that memories of such times may account for our stories of giants and "little people." In the same spirit, I would like to toss out the speculation that the story of Peter Pan and other legends of eternal youth are linked with some deep inner knowledge of our own immaturity.
† The experts don't always like admitting how little we know even about *Ramapithecus*—just a few fossil fragments, identified as an ancestor of the human line from the jaw shape. It *could* just be wishful thinking, and that might resolve all the dating problems at a stroke!

line of tree-apes standing, metaphorically, at the edge of the woods gazing at the open plains at 10 Myr BP, and then the film cuts immediately to a variety of much more human hominids populating East Africa 3 million years ago, working with stone, hunting animals for their meat, sitting around the camp carving wood into tools, and so on. *Ramapithecus* lived in Europe and Asia as well as Africa, but the evolutionary pressures which produced such a diversity of hominids seem to have occurred only in Africa, for reasons we simply do not understand. All we can do is take up the threads of the story of human origins on the other side of that gap, in Africa, where the last of the direct *Ramapithecus* line lived alongside the early forms of our own line (*Homo*) and two other hominid lines, *Australopithecus africanus* and *Australopithecus boisei*, which were related to the *Homo* line but do not seem to have been our direct ancestors. The story of human evolution from then on is essentially one of how this diversity of hominids was whittled down to just one, ourselves. That is the story which has been transformed in the past two decades by fossil discoveries made in East Africa, and which is improved by new discoveries almost every year—one of the great scientific developments of the twentieth century.

The neatness of the labels attached by the experts to fossil remains of different human and near-human species, and the long gaps in the fossil record, tend to conceal the subtlety of the evolutionary changes in our ancestors. If we had a complete skeleton from every generation stretching back from me or you to, say, *Ramapithecus*, there would be nowhere along the long line of ancestral bones where we could pick out a shift from *Ramapithecus* to *Homo habilis*, *Homo habilis* to *Homo erectus*, or *Homo erectus* to *Homo sapiens*, modern man. But bones from 500,000 to 1 million years ago would be clearly different from present-day bones, and from remains dated at 3 Myr BP or 15 Myr BP. These broad differences between remains differing in time by hundreds of thousands, or millions, of years are the basis for the different classifications; but nowhere along the evolutionary path could you "spot the join." There is no suggestion that with one sudden mutation *Homo habilis* gave birth to *Homo erectus*, and indeed this direct line of descent is still not proven as the line of human origins. *Homo habilis*, judging by the remains from 3 Myr BP, looks like a descendant of *Ramapithecus* or of something like *Ramapithecus*, a ramapithecene. The two "cousins," also seemingly descended from ramapithecene forebears, were also around between 2 and 3 Myr ago, *Australopithecus africanus* and *Australopithecus boisei*. These were also once thought to be part of the main line of human evolution, and although the best evidence now available indicates that they were side branches rather than our own direct ancestors, the issue is not entirely clear-cut. What is clear, though, is that several manlike creatures existed around 3 million years ago, the products of a successful line of evolution which was spreading into different ecological niches and diversifying. In the course of the past 2 million years, there have been many extinctions of species on Earth, including many species of mammals and all but one of the manlike species. Unlike the extinctions which brought the end of the era of the dinosaurs, 65 Myr ago, there is no mystery about this latest wave of extinctions. It has coincided precisely with a great wave of Ice Ages, especially across the Northern Hemisphere, which have been directly brought about by the slow shift of continental masses into high latitudes around a landlocked polar sea, and which have been modulated during the past couple of million years by the regular rhythms of the Milankovich cycles, which bring a repeated ebb and flow of ice across the northern lands.

ICE AGE EXTINCTIONS

The extinctions associated with the repeating Ice Ages of the past few million years—an Ice Epoch—mark the end of the Pleistocene. The known story of human evolution in the past couple of million years, then, is one of adaptation under conditions of unusual evolutionary stress, with several closely related species dying out. The unknown story of prehuman evolution in the long fossil gap before about 3 million years ago must have produced first a diversification of the Ramapithecene stock to produce a pre-*Homo* and a pre-*Australopithecus*, with *Australopithecus* later dividing into the two forms found in the 3 Myr-old remains, *africanus* and *boisei*.*

It is only in the past decade or so, however, that the story of human evolution over the past couple of million years can confidently be described as "known." Since 1972, a succession of finds in the East African sites has produced what seems to be a definite lineage, in spite of the gaps in the fossil record. One crucial discovery was a skull of *Homo erectus* some 1.5 million years old— *erectus* was the immediate ancestor of *Homo sapiens*, and his remains have been found in many parts of the world, across Europe and Asia, where he lived up until about 500,000 years ago. The discovery of such an old *erectus* skull in East Africa bears out the supposition that the earlier form, *Homo habilis*, became *Homo erectus* in East Africa and the species then spread (under the drive of natural curiosity) out into Europe and Asia. Another particularly interesting find is a skull roughly half a million years old which is, in some respects, intermediate between *erectus* and ourselves, *Homo sapiens*. The argument runs that *erectus* came out of East Africa to spread over the world, and that the continuing evolutionary pressure encouraging the development of the best assets for survival then carried *erectus* on to become *Homo sapiens* everywhere he went—something like the way in which both Old World and New World monkeys evolved independently from the same original stock.

It might seem odd, in that case, that *erectus* did not diversify more. But he did! From about 100,000 years ago a slightly different form of *Homo sapiens* is found in fossil remains across Europe, in China, and in other parts of the world. This was a kind of man adapted, in some ways, more effectively for the harsh Ice Age conditions—Neanderthal man, now called *Homo sapiens neanderthalensis*, while our own line is most accurately dubbed

---

* Dating of the East African sites, incidentally, is made a lot easier because there were many volcanoes active in the region during the period of interest spanning the past few million years. Layers of volcanic ash in sediments can be dated accurately using radioactive isotopes and other techniques, and a fossil bone found in sediments between two layers of volcanic ash must have been buried between those two well-dated eruptions! It is intriguing to speculate that the special conditions associated with the volcanic activity and the creation of the great rift valley of East Africa might have been part of the pattern of evolutionary stress which caused the emergence of modern man—but that is only speculation.

Homo sapiens sapiens. The Neanderthals had a bad press when their remains were first found in the mid-nineteenth century. They were the first obviously archaic human form to be found, and this led naturally, at that time, to the assumption that Neanderthal was an "ape-man," while the unfortunate coincidence that one almost complete skeleton found in southern France happened, we now believe, to be the remains of an old and arthritic individual bore out the assumption that Neanderthal man was a nasty, stooping, brutish sort of person. The story revealed by modern analysis of a much greater variety of remains is far from that. The Neanderthals, the experts now tell us, were contemporaries of *Homo sapiens sapiens,* not "ape-man" ancestors; true, they had thicker skulls giving a "beetle-browed" appearance, and were stockier than our own line. But they actually had larger brains than ours and lived complex lives during extremely harsh conditions in northern Europe. And from the height of the latest Ice Age, over in the mountains of Iraq the remains of a cave burial, some 60,000 years old, have been uncovered. The individual laid to rest in the cave was surrounded by masses of flowers from the nearby fields, revealed today by the pollen preserved on the cave floor. This is a fascinating insight into life at the time in itself— but doubly so since many of the species of flower used in the burial are traditional herbal medicines. The people who carried out this burial, surrounding their dead friend or relation with flowers, were clearly "human" in a very real sense, and they may have had a knowledge of herbal medicine. And in case you haven't guessed, they too were Neanderthals, not *Homo sapiens sapiens.*

What happened to the Neanderthals? The youngest of their fossil remains are dated at around 30,000 years. Perhaps they could not, in the end, compete successfully for resources with the more efficient *Homo sapiens sapiens.* Although there is no suggestion that there was any direct conflict between the two, if our ancestors ate the best food then the Neanderthals had to go without. Whatever did happen to them may have been related to the warming at the end of the latest Ice Age, since their remains are found mainly in the north. In all probability, though, they were not "wiped out" in the normal sense of the word. Neanderthal and what we would

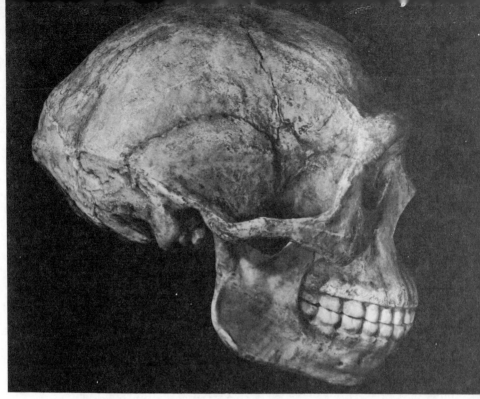

Skull of the immediate ancestor to *Homo sapiens, Homo erectus.*

Skull of our nearest extinct relative, *Homo sapiens neanderthalis.* © *British Museum (Natural History)*

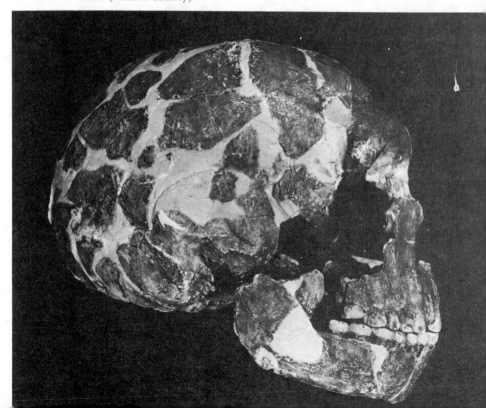

call human stock had scarcely had time to divert much from one another, and could certainly still interbreed. The differences between the two were small, and may have been regarded as no more significant than tribal differences. Neanderthals may well have interbred with *Homo sapiens sapiens*, so that Neanderthal genes, somewhat diluted, may still survive in the human gene pool today. I hope so—I've no objection to being related to the people who carried out a burial in the Shanidar Cave of the Zagros Mountains one June day 60,000 years ago.

MODERN MAN

In genetic terms, the story of evolution is virtually up to date with the emergence of *Homo sapiens* half a million years ago and the short-lived Neanderthal experiment. The story of how *Homo sapiens sapiens*, honed by the succession of recent Ice Ages and associated changes in the forests and grasslands of the world, went on to develop civilization would require another shift in focus, one so extreme that it could only be dealt with in another book. Fortunately, the job has been done many times,* and I can, with a clear conscience, leave the story of human origins with the end of the Neanderthal experiment. From then on, with the end of the most recent Ice Age, it was just a question of using the skills developed by evolution; the critical invention, some 10,000 years ago, was agriculture, which led to a settled way of life, to trade, villages, cities, and, it must be faced, to warfare. Before I leave the story entirely, however, I want to pick up on this last point, human aggression, since so much confusion has been spread regarding our "terrible hunting instincts" and "uncontrollable fighting urges." The truth is our hunting ancestors led much more peaceful lives than the first farmers—the blame for warfare can be laid at the door of agriculture and the concept of property, not on the shoulders of ancestral hunters!

The ancient way of life, the hunter-gatherer tribe, is preserved today in several parts of the world, most notably among the !Kung who live on the northern fringes of the Kalahari Desert. By study-

*The best is surely Jacob Bronowski's *The Ascent of Man* (London: BBC, 1973).

ing such people who today make a living off the land under harsh conditions, anthropologists can get some idea of how our ancestors lived. The first surprise is that, even on the fringes of the Kalahari, the living is easy. Even the term "hunter-gatherer" is inappropriate, since most of the food comes from gathering, and it has been suggested that "gatherer-hunter" is more appropriate. The !Kung mainly eat mongongo nuts, found in profusion in that part of the world, to the tune of 1,260 calories and fifty-six grams of protein a day, and they see no need to plant crops when nature provides such a bounty for the picking. They also eat fruit, berries, and so on, and meat from various animals and birds. Hunting is men's work—but it involves each man in hunting for only about nineteen hours a week. Gathering is women's work, and takes them just a few hours a day.

In this particular case, the easy life rests upon the availability of the mongongo nut—but though the specific staple food may vary, the success of the gatherer-hunter way of life in many parts of the world shows a similar, typical pattern. After all, if this were a precarious way of life, our ancestors would never have survived as they did for millions of years before the invention of agriculture! There *is* a key to such a way of life, though, and it is closely connected with the low birthrate, which maintains a stable population in tune with the available resources, in this case the number of mongongo nuts.

The !Kung women have babies on the average only every four years, so that each woman produces only four or five offspring during her fertile life. With half the infants surviving to adulthood, a stable population is assured. How does this come about? Partly it seem to be connected with the long time for which infants are suckled, contributing a natural contraceptive effect.* But partly it is because infanticide, especially of girl babies, is an accepted, if hidden, part of gatherer-hunter life. In tune with their environment, the gatherer-hunters know that a baby born "too soon" after a sibling cannot be reared without the sibling suffering and probably dying. Only when agriculture began to produce surplus food

---

* One recent suggestion links the effectiveness of this natural contraceptive to the frequency of suckling. "Little and often" meals for the baby seem to release a constant stream of the appropriate hormone in the mother.

Members of the !Kung tribe. Studies of hunter/gatherers such as the !Kung have exploded the myth that our ancestors lived by hunting and had evolved a "killer instinct." These tribes, living in regions of the world regarded as marginal for human life in modern terms today, actually base their existence on gathering available food such as nuts, and don't even have to work very hard at that. Hunting is an occasional occupation, not a way of life; we are descended not from a line of killers, but from a line of gatherers making the most of what food came their way, with little thought for the future. Most of the problems of violence associated with modern life—warfare, theft and so on—can be linked directly to the invention of agriculture, the need both to plan for the future and to store the harvest, and the concept of property. Farming, it seems, leads to killing; our ancestors—and the !Kung today—are best described not as hunter/gatherers but as gatherer hunters. © *British Museum* (*Natural History*)

could our ancestors successfully rear large numbers of children to adulthood—and that began the population explosion.

This understanding of the gatherer-hunter way of life immediately pulls the rug out from under theorists such as Robert Ardrey, who argue that our ancestors' need to hunt in order to live provided the evolutionary pressure which makes us all potential killers today. In fact, all the evidence suggests that our ancestors were gatherers who did a bit of hunting on the side, and had no need to develop a "killer instinct" since there was plenty of food around. Richard Leakey and Roger Lewin, in their books *Origins* and *People of the Lake*, convincingly make the case that neither hunting nor gathering was the key to human origins, but that the combination of the two, a gatherer-hunter mixed economy, is what made our ancestors so special. As for bloody conflict between members of the same species, for man as for any other animal this is a rapid road to evolutionary failure. Genes for fighting to the death don't survive well, because the bodies in which they reside are all too often fought to death; "successful" genes are the ones which provide for bluff and mock battle in the case of disputes (over a female or territory, perhaps) but acknowledge defeat early and live to bluff another day. And it is no use arguing, as even so respected an authority as Desmond Morris does in his book *The Naked Ape*, that our "bluff" signals were no longer adequate when our ancestors first picked up rocks and used them as weapons, so that what should have been a threat display turned into bloody mayhem. If weapons were used in conflicts between our ancestors 5, 6, or 10 million years ago, then the laws of genetics have had ample time to operate. Our ancestors who had a lust for battle died in battle, leaving survivors with less battle lust to reproduce and produce the next generation.

AGRICULTURE AND AGGRESSION

No, the source of our present aggressive behavior and predilection for warfare must be sought much closer in time. The argument that our natural "signals" have been overwhelmed is a good one, but it can only be applied to an event so recent that evolutionary adaptation to the changed situation has not yet happened. The only candidate is the invention of agriculture, a mere 10,000 years ago. As I have already pointed out, this led to the population explosion, and there can be no argument about the fact that we are not evolved to live in such large numbers occupying small spaces, as Desmond Morris himself stresses in another book, *The Human Zoo*. Even more to the point, though, agriculture provided something new to fight over. First, farming requires that the farmer stay with the land throughout the year, instead of wandering about in search of food. So communities develop, with stockpiles of food and fields full of crops. Just as in the dove/hawk scenario, it is now clearly an advantage for someone who has not gone to all the trouble of growing the crop to come in and steal it, while the farmer has to fight back or starve. The difference, 10,000 years ago and since, is that the situation depends not on inbuilt genetic programming, but on the intelligent ability of our species to reason out an advantageous course of action. And since that time there has always been conflict over food, land, property, and "living space." Possessions encourage conflict, and while property may not be theft, the aphorism "property is war" seems quite appropriate.

Just as in the hawks/doves scenario, everybody could be better off by cooperating, but the prospect of advantage for a minority— say, one country—to be gained by war is so great that someone— some country—is almost bound to try it. And this, not our mythical hunting ancestry, is the cause of human conflict. The difference, again, is that we are aware of the situation and are *not* being driven by the blind instructions of our genes. And it is important to recognize this. If we are programmed to fight, it is difficult to see how the ultimate conflict could be avoided; but if we are programmed for the quiet gathering life, and have merely been confused by recent developments, then there is every reason both to seek and to expect to find an intelligent solution to the problems of inequality in the world today. If we fail, and a nuclear holocaust does destroy civilization, this will not indicate that *intelligence* is necessarily a "failure" in evolutionary terms, since the !Kung, among others, certainly use their intelligence to maintain their way of life. It might indicate that agriculture was a mistake, which is an interesting thought. But looking at the other side of the coin, if, in spite of the agricultural mistake, we can avoid disaster, then that will indicate that intelligence really is an extremely useful evolutionary factor. The evidence that the population explosion is just beginning to come under control hints that there is at least a glimmer of hope. Maybe we will outlast the dinosaurs yet; maybe we will even take life off the Earth and back into space, from which the life molecules first came to Earth. Skipping just the past 10,000 years of history in the story of where we come *from*, there remains only the prospect of where we are going to. The origins of man and the Universe seem remarkably well understood, but what is our destination?

9 DESTINATIONS

The future of man and the Universe can be viewed on the same diversity of timescales as our origins. On the shortest of timescales, the next 50 to 100 years or so, the future is entirely in our own hands. During the 1970s, a wave of gloomy prognostications about the future appeared, many of them building on the theme that our planet is already overpopulated and that it will be impossible to feed the population of the early twenty-first century. A few futurologists, notably the Hudson Institute group headed by Herman Kahn, have produced forecasts which seem superficially like wild science fiction stories, visions of plenty for all just around the next corner as supertechnology at last removes all drudgery from human life. And in the early 1980s, as the first hints of supertechnology have appeared with microprocessors, the gloom-mongers talk of mass unemployment, while the optimists talk of more "leisure time," as robots take over the role of slaves. The fact that both optimists and pessimists can make out convincing cases, however, shows how wide the choice facing us is. Either view could be right, depending on what we do now; and most probably, of course, the immediate future of our own lifetimes will produce neither the collapse of civilization nor the advent of a new paradise on Earth. We will surely muddle through with some people—some nations, and blocs of nations—doing very well and others doing much less well.

This diversity between rich and poor, haves and have-nots on a global scale, received welcome publicity in 1980 with the publication of the report of the Brandt Commission on the split between North and South in global terms. The publicity was due not so much to anything startlingly original in the report as to the fact that it had appeared in the names of such respected politicians as Willy Brandt from West Germany and Edward Heath from the United Kingdom. The eighteen members of the commission, from as many countries' scattered across five continents, came up with recommendations for the solution of the major problems facing the world now that can be summed up in one word: "Equality." Tension and problems will continue to exist on a global scale as long as there is such a huge difference between rich and poor; although the Brandt Commission does not explicitly make the point, we are still facing the problems of possessions that began with the invention of agriculture! In simplistic, idealistic terms, if the resources of the Earth were managed wisely and divided fairly among all nations, and all individuals within nations, the causes of conflict would disappear, along with poverty, starvation, and ill health. The commission's report goes further than some such prognostications by spelling out how this might be achieved in political terms. In the terms I have used in this book the problem is one of maintaining a population of "all doves" in which everyone has a fair share. Such a situation may well provide the best opportunity for every member of the population, but there is initially an enormous "evolutionary" advantage for any hawk which tries to grab more than its share. If genetics alone were involved, there would be no way in which an "all-dove" population could remain stable. But we are no longer dealing with evolution of the human species in genetic terms; we have the ability and intelligence to consider alternative courses of action and to calculate their outcomes before choosing the most beneficial. We are now faced with the biggest test of our kind of intelligence, perhaps the most crucial watershed in human development since the invention of agriculture. If we follow the best course of action for the majority, then after a few difficult decades our future should be assured, and even the science fiction–like predictions of the Hudson Institute may not prove so wide of the mark. The alternative is that a minority, the hawks, will use their intelligence and foresight to grab as much as possible for themselves. This is the course that has been followed for all of recorded history, and the "hawks" in these terms include such countries as Britain, the United States, Japan, all European countries, the USSR, and the Eastern Bloc. There aren't really any "doves" in the world today, since the have-not nations are simply trying to follow the same path that the present-day rich nations have already followed. The prospect is hardly encouraging, but the Brandt Commission report may be a significant chink of light in the gloom.

GROWTH FOR SURVIVAL

One thing that is very clear, and cannot be overemphasized, is that the physical limits to the "carrying capacity" of planet Earth have not yet been reached. I was among the many people persuaded by the loud noises of the doomsday brigade that the end really was nigh in the early 1970s, and as a result I spent three years with the Science Policy Research Unit (SPRU) at the University of Sussex, working on the fringes of a team studying world futures in general and the problems of food, energy, and raw materials in particular. Such studies show clearly that it would be easy to feed adequately even twice the present population of the world, using no more than existing technology and skills correctly applied on existing farmland. People starve in the world today because they do not have money (at the individual or national level) to buy food on the world markets—a damning indictment of the present economic system. With only a fraction of the effort now devoted to "defense" diverted into satisfying the needs of the hungry, starvation could be eradicated from the world and the population stabilized. Nor are the problems of energy and raw materials insurmountable, although, clearly, food should have priority. And there are already clear signs that the population explosion is slowing down, and that the days of exponential growth—the prospect which so frightened the doomsayers of ten years ago—are already behind us.*

In terms of the broad sweep of the Universe, then, there is a real prospect that the human race will survive for long enough to be interested in the changes going on as a result of natural processes both on our own planet and outside. *If* we do not bring ruin upon ourselves, then our descendants may be around to witness, and perhaps interfere with, natural processes on a timescale of tens and hundreds of thousands of years (most notably ice ages) and perhaps even millions of years (where the effects of continental drift and long-term tectonic activity begin to be noticeable). The most immediate natural hazard facing "civilization as we know it" is indeed the arrival of the next full ice age, which is now well due by geological standards. Remember that with the present arrangement of the continents the Milankovich cycles dominate the rhythms of ice ages, and remember also that in round terms the "interglacials" between full ice ages are roughly 10,000 years long. It is now a good 10,000 years since the most recent ice age waned, and projecting forward the interlocking rhythms of the Milankovich cycles, we can come up with a confident forecast that, other things being equal, the world will get colder for about the next 5,000 years, sliding into a new ice age somewhere along the way. The pattern then would be for ice to maintain its grip over the northern land masses for roughly 100,000 years, and it is a nice touch of irony that the countries which would be worst affected are almost exactly those countries which are today the "rich North." They— we—perhaps deserve such a fate. But it is not really likely to happen.

This time around the Milankovich cycles, "other things" are not equal and there is a new factor to take into account. Human activities are already significantly altering the composition of the atmosphere of the Earth, adding carbon dioxide as fossil fuel is burned

---

* I reported the findings of the SPRU team in my own book *Future Worlds;* because the "limits to growth" debate is still seen by many people, mistakenly, as synonymous with the "futures debate," I have included here, as Appendix B, an article on the topic which originally appeared in *New Scientist.* The most comprehensive account of the whole futures debate is *World Futures,* edited by Marie Jahoda and Christopher Freeman, published in the United Kingdom by Martin Robertson and in the United States by Universe.

and tropical forests are destroyed. The effect of this is likely to be a small but significant warming of our planet, through an enhancement of the greenhouse effect. This in itself may or may not be enough to hold off the "next" ice age. But there can be little doubt that if we survive for another thousand years or so to produce a genuine global society in which resources are concentrated on real problems, then our descendants will have the ability to manipulate the environment to the extent of preventing the spread of ice—if, of course, they want to.

GAIA AND MAN

A hint of what may be a good reason not to upset the global applecart by any such interference with the natural rhythms has come from Jim Lovelock, a British researcher who has developed what is now known as the "Gaia hypothesis," taking conventional ideas of the ecological balance between different kinds of living organisms on Earth to the seemingly logical conclusion that over thousands of millions of years life on Earth has helped to maintain a stable situation with conditions well suited to life. In spite of the great extinctions that have occurred from time to time since life first got a grip on Earth and converted the atmosphere to an oxygen-rich state, there have been very few fluctuations in temperature, atmospheric composition, and so on. Could such stability really have been simply good luck? Lovelock argues that this is unlikely, and points to the aweful examples of Venus and Mars as planets left to their own devices, with no living regulators to maintain this kind of stability.

Working with Lynn Margulis, who was herself a pioneer of the idea that modern cells are built up from different kinds of early, simple cells that learned to get together in a cooperative venture, Lovelock has been able to explain, in principle, how the stability of the present atmospheric system is maintained by a variety of checks and balances involving such unlikely organisms as the bacteria which produce marsh gas, the peat bogs which lock up carbon and thus prevent the concentration of carbon dioxide in the atmosphere from rising, and so on. The idea of an almost sentient single Earth-creature (dubbed Gaia after the Greek Earth-goddess) is so bizarre that Lovelock's hypothesis is regarded with suspicion by most scientists. It is, however, scarcely any less likely that many different forms of life on Earth may be unconsciously maintaining the stability of the whole environment through a variety of checks and balances than that thousands of millions of individual cells, each of them a combination of earlier cell forms working in harmony, could cooperate to produce a human being, maintaining stability of body temperature, fighting off infection, and even providing self-awareness in an intelligent brain. Using language which would make a geneticist wince, we can even speculate that mankind was "invented" by Gaia to serve as her own intelligent brain and nervous system, overseeing the whole planet and taking care of problems that cannot be solved so easily by unconscious feedback. Unfortunately for Gaia, at present we are doing more harm than good. Maybe the best course of action would be to plant trees on an enormous scale to take carbon dioxide out of the air again, and to let the ice ages follow their agelong rhythm while we bend with them and learn to adapt to the needs of Gaia rather than destroying natural balance to suit our short-term convenience. Lovelock and Margulis are not worried that we might upset the balance so much that Gaia could not survive: long before that point was reached, conditions would be altered so drastically that our species—at least, the "civilized" members of it—would be wiped out, and Gaia could then rebuild, as after previous crises such as the events associated with the great faunal extinctions of some 225 million years ago. It is for our *own* sake, according to the Gaia hypothesis, that we should work with, rather than against, the natural systems which maintain the comfortable environment on our planet. One of the examples of our role as "nervous system and brain" for Gaia, an example chosen by Lovelock himself, dovetails neatly with the fact of previous great faunal extinctions and one of the ideas spelled out early in Chapter 8. Someday, as Lovelock points out, a large asteroid must collide with the Earth, just as large asteroids have done in the past:

> The potential damage from such a collision could be severe, even for Gaia . . . with our present technology, it is just possible that we could save ourselves and our planet from disaster

. . . by using some of our store of hydrogen bombs and large rocket vehicles . . . to convert a direct hit into a near miss.*

So the potential value for Gaia of a species like ourselves is clear, although Lovelock does not seem to be aware of the irony that it may have been precisely the catastrophe he cites, the impact of a large asteroid, which wiped out a previous candidate for the role of Gaia's nervous system, *Saurornithoides*, 65 million years ago!

THE ANTHROPIC UNIVERSE

However far you choose to go along with the Gaia hypothesis, there is no doubt that we are creatures of the Earth, evolved and adapted to conditions on Earth. It is equally true, though, that we are creatures of the Universe in which we live, evolved and adapted to fit our surroundings on a cosmic scale, although this may seem much less obvious at first sight. Indeed, some people find it so unobvious that the suitability of our Universe for life as we know it is sometimes raised as a serious philosophical or theological puzzle. Isn't it remarkable, runs this kind of argument, that the Sun is just the right temperature to keep the Earth warm, and that it stays just the right temperature long enough for life to evolve? Isn't it remarkable, indeed, that all the processes since the Big Bang have been just right to produce us—the way galaxies and stars formed, the way planets formed, even such fundamentals as the strength of the gravitational force, seem to have been tailored precisely to the needs of people on Earth! In fact, of course, what has happened is that *we* have been tailored to fit our surroundings, including the strength of the gravitational force and the temperature of the Sun. Our form of life depends, in delicate and subtle ways, on several apparent "coincidences" in the fundamental laws of nature which make the Universe tick. Without those coincidences, we would not be here to puzzle over the problem of their existence—so, by definition, any universe in which we exist to do the puzzling must provide those puzzles to think about. Cosmologists can construct models of universes with different physical

* J. E. Lovelock, *Gaia: A New Look at Life on Earth* (London: Oxford University Press, 1979), p. 147.

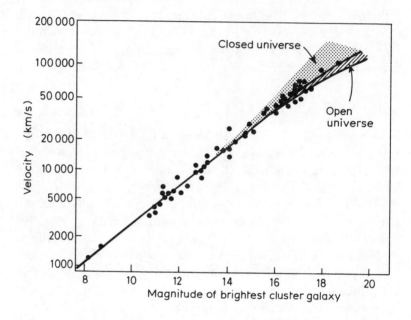

Is the Universe open or closed? Its ultimate fate depends on the density of matter within it, which is very difficult to estimate. A direct test of how quickly the expansion of the Universe is slowing down would answer the question once and for all, and in principle this could be done by comparing the red shift of distant galaxies, a measure of their recession velocities, with brightness (or rather, dimness!), an indicator of distance (note that by convention the magnitudes by which astronomers measure brightness are *bigger* for fainter objects). But in the observable Universe, these measurements (blobs) are inadequate to distinguish between the range of values corresponding to an open Universe, destined to expand forever, and a closed Universe, destined ultimately to collapse into a fireball reminiscent of the Big Bang itself. All we can say is that the real Universe lies very close to the dividing line between open and closed.

laws, and it may be that such universes exist or have existed (everyday intuition about simultaneity rather breaks down when pondering when or where another universe might exist). The "anthropic principle" says that our Universe seems to be tailor-made for us because people like us can only evolve in this kind of universe; and Barnard Carr and Martin Rees spelled out just what this means in a detailed article published in *Nature* in 1979.

Without going into all of those details, some of the "coincidences" relevant to life in our Universe can be spelled out in simple terms. Going right back to the beginning of the Universe in the Big Bang, when about 20 percent of the original hydrogen was converted into helium in the fireball, calculations show that only a very small change in one of the constants which determine the rate of nucleosynthesis—the fine structure constant—would have led to a completely different kind of universe, containing either very little helium or very little hydrogen. Such a situation—either extreme—could never have produced the sort of stars which we see as typical in our Universe, and without such stars the story of life would have been very different, if life could have evolved at all. A rather less abstract "coincidence" concerns the relative sizes of things in the Universe. One way of measuring the proportions of an object on the human scale in comparison with the size of a molecule or a galaxy is to use the "geometric mean," multiplying the big and small numbers together and taking the square root to find out what a "typical" number between the two extremes ought to be. This method gives a genuine scale of sizes for comparing the very big with the very small, and removes some of the problems involved in making such a comparison of opposites on an everyday linear scale, where you might add the big and small numbers together, then divide by two, to give the arithmetic mean.

It turns out that the size of a planet like Earth is the geometric mean of the size of the Universe and the size of an atom, and that the mass of a man is the geometric mean of the mass of the planet and the mass of a proton. These relationships *might* just be coincidence. But on the other hand, there are very good reasons, based upon the actual values of the electric and gravitational forces in our Universe, why a manlike creature on a planet like the Earth should be just about our size.

Provided an animal is successful in evolutionary terms and comes to dominate its environment, there is generally evolutionary pressure toward an increase in size. Bigger individuals in each generation may get more food or may find it easier to find mates, provided that they are not bothered by predators and become so big that they cannot easily hide. This tendency shows up, for example, in the way small dinosaur species radiated after each extinction of the larger dinosaurs and diversified to produce new, large species. Over the past 65 Myr, small mammals have moved into empty dinosaur niches and grown in the same way. From the point of view of a physicist, though, it is easy to calculate how big a human being can be and still survive on our planet. Our bodies are held together by electric forces, the forces which bind atoms together as molecules and enable molecules to hold onto each other. Working against this is the force of gravity, which will break apart any collection of molecules that is too big if it falls over— whether the collection of molecules is a tree, a rock, or a man makes no difference to the law of gravity. If it is bigger than a certain size, it is sure to break if it falls down. So both the electric forces which dominate the structure of atoms and the gravitational forces which hold planets together are important in man—and it is, therefore, no surprise to find man intermediate in size between an atom and a planet!*

Looking again at the bigger scale, if the constant of gravity were a little bigger than it is then the convection inside stars would be suppressed and the instabilities which produce supernova explosions could never develop, so heavy elements would never be scattered across interstellar space to form planets at all. And so the list goes on; everything about us can be interpreted in a very precise way as the result of a very exact "choice" of physical laws and constants of nature. The conclusion reached by many people is that a

---

* In case you are wondering, as I did when I first came across this argument, man is indeed one of the largest animals on Earth today. The argument doesn't apply to whales, of course, since they are supported by the water in which they swim, and the lumbering progress of an elephant shows how gravity begins to dominate for animals bigger than man: the tiger, and so on. Only a slight refinement is needed to complete the story outlined above: on a planet like Earth, no *agile* creature can exceed the size of a man and live!

universe suitable for life is a very unusual phenomenon. In the mathematical equations, it is quite possible to vary a tiny detail, such as the rate at which hydrogen is converted to helium in the fireball, or the constant of gravity. Such changes produce new equations describing entirely plausible universes, but almost always ones in which life as we know it is impossible.

What does this mean? One possibility is that the Universe we know is a highly improbable accident, "just one of those things." Alternatively, it may be that everything we think of as "the Universe" is just a small corner of something much bigger, and that in most places away from our corner there are different physical laws and no chance of life. The possibility which I like best goes one step further even than this, suggesting that there may be a literally infinite number of alternative universes which exist in some sense "side by side" in time or which follow their life cycles one after the other. There is no observational evidence for or against such an idea, and probably there never will be. But it would mean that every kind of possible combination of physical circumstances happened someplace or sometime. In such a situation, life could only exist to puzzle over the mystery of the Universe (or universes) in those very rare places where the very unlikely conditions necessary for life just happen to occur. Life must exist, by this definition, in a very unlikely Universe so puzzling as to make intelligent life baffled as to why it should exist at all!

THE FATE OF THE UNIVERSE

It is remarkable enough that life on Earth should have evolved to the stage where we are able to puzzle over such questions, with the Sun as yet only halfway through its life as a reasonably stable, main-sequence star. There is still another 5,000 million years or so before our Sun comes to the end of its main-sequence life and expands as a red giant to engulf the inner planets, and there is no reason to expect the extinction of life on Earth—the death of Gaia—before that time. There is, though, no reason to expect human life to survive for even a fraction of that time. We may be victims of a natural disaster—a nearby supernova explosion, perhaps, producing a flood of radiation and massive faunal extinctions across the Earth—or of our own folly. If so, given the events of the past 65 million years, Gaia will have ample time to try again to produce an intelligent "nervous system" from some other species. We may be just the first of many intelligent forms of life to develop here. On the other hand, we have developed something new in the history of the Earth, and may change circumstances so much that the story of evolution here is never quite the same again. We store information outside our own brains, in books, libraries, and now computer systems, and this represents something beyond biological intelligence. Richard Dawkins in *The Selfish Gene* goes so far as to describe units of information—ideas expressed in words, a song or a book or a play—as "the new replicators," dubbing them "memes":

> Just as genes propagate themselves in the gene pool by leaping from body to body via sperms or eggs, so memes propagate themselves in the meme pool by leaping from brain to brain . . . if a scientist hears, or reads about, a good idea, he passes it on to his colleagues and students.*

Dawkins even goes on to consider competition between memes for available space in books, computer systems, or even in human memories. The analogy is imperfect when pushed to extremes. But it does clearly show the emergence of something different in our culture which, barring overwhelming catastrophe, may make the culture, or its information content, more durable than the genes of, say, the large dinosaurs. However we may change, or even if we are "superseded" by computers and robots as in some science fiction scenarios, the memes of human culture—and of human individuals such as Einstein or Shakespeare—may persist. And if we ever establish communication with other intelligent species on other planets circling other stars, then human memes may spread across the Universe even if physical interstellar space travel remains impossible.

Perhaps, though, our descendants will develop a space culture—if not an interstellar one, then a truly interplanetary one—

---

* Richard Dawkins, *The Selfish Gene* (London: Oxford University Press, 1976), p. 214.

using the resources of the whole Solar System rather than one small planet. But however optimistic you may be about the immediate human prospect, it is inconceivable to imagine mankind and human society persisting in any recognizable form for even a thousand million years when so much has happened in the past thousand million years. We are only temporary visitors on any truly universal timescale, and the question of the ultimate fate of the Universe, long after the human race has passed away and the Sun is dead, can only be of philosophical, not practical, importance. But that doesn't mean that it isn't interesting.

The key factors which determine the ultimate fate of the Universe are the amount of matter it contains and the rate at which it is expanding. The matter tends to slow down the expansion through the action of gravity, trying to pull the Universe back into a compact state reminiscent of the fireball of the Big Bang. In simple terms, the Universe can only expand forever if it is exploding faster than the "escape velocity" from itself. The rate of expansion—the value of the Hubble constant—varies as the Universe evolves, so that it might seem difficult to decide here and now what the fate of the Universe will be in many thousands of millions of years. But as the Universe expands its density decreases as matter is spread more thinly across space, and this decrease in density is exactly in line with the rate at which the Hubble constant changes. If the density of matter across the visible Universe we see today is sufficient to halt the expansion we can observe today, then the Universe has always been exploding at less than its own escape velocity, and must eventually be slowed down so much that the expansion is first halted and then converted into collapse. On the other hand, if the expansion we observe today is proceeding fast enough to escape from the gravitational clutches of the matter we observe today, then the Universe is and always was "open" and will expand forever. "All" that astronomers have to do is measure the density of matter across the known Universe and the present-day value of the Hubble constant which tells us the rate of universal expansion. Then they will be able to answer unambiguously the question of whether our Universe is open and will expand forever or closed and must ultimately collapse.

Neither of those two crucial measurements is easy, and the situation isn't helped by the fact that the Universe we live in seems to be rather young in terms of its expansion. During the time since the Big Bang, the Universe has changed very rapidly, and has yet to settle down into a quiet middle age—looking back to the time when life first emerged on Earth, the density of matter in the Universe must have been twice what it is today, and going back roughly the same distance again into the past all the matter in the Universe was piled up in one place in the Big Bang.* The progress from Big Bang to ourselves has been spectacularly swift when measured against the expansion of the Universe itself, and the future evolution of the Universe stretches out much further ahead than its history since the Big Bang, whether it is open or closed.

The youth of the Universe makes a whole family of tests of its ultimate fate useless. In principle, it would be possible to find out how rapidly the expansion of the Universe is slowing down by comparing the recession velocities of nearby objects, which are seen by light which left them recently, and distant objects, seen as they were when the Universe was young. This would show how the Hubble constant has changed, and whether the change is sufficient to imply that the Universe is closed. This is a very neat idea, and removes the problem of measuring the density of the Universe. But all that these particular observations can tell us is that the rate at which the Universe is slowing down is close to the critical value which determines whether it is open or closed. Because the value of this "deceleration parameter" is close to the critical value, a long span of the history of the Universe must unfold before it is possible to use this test to decide the issue one way or the other, since small differences in the "here-and-now" and "there-and-then" measures of the Hubble constant are crucial. And enough of the history of the Universe hasn't unfolded yet for the observations to give a completely clear-cut answer.

This doesn't stop people from trying to pull an answer out of the hat, though, and some cosmologists do argue that the deceleration

---

* We can't say that the Universe was only half as big as it is now when life appeared on Earth, since the Universe may actually be infinite, in which case it always was infinite even when it had very high density. But by saying the density was twice the present value a few thousand million years ago, astronomers do imply that a particular galaxy or group of galaxies now visible in the sky must have been that much closer to us at that time.

parameter is now known accurately enough to settle the issue. Lloyd Motz, professor of astronomy at Columbia University, is a strong supporter of the closed universe model, and he cites measurements of the deceleration parameter made by American observer Allan Sandage as conclusive evidence which

> shows that the galaxies are receding at speeds less than the speed of escape. The expansion of the Universe must ultimately stop.*

This seems a rather strong line to take just yet in the light of the available evidence, and most astronomers would not agree with Motz's firm conclusion. Instead they turn their attention to the other test of the ultimate fate of the Universe, which depends on measuring the expansion rate (which can be done reasonably accurately) and comparing it with the density of matter in the Universe (which is very difficult to assess).

The amount of matter required to "close" the Universe by its overall gravitational influence is small in density terms—just one gram in each cubic volume of space measuring 35,000 km along its sides. The matter we can see in the Universe is almost all in the form of bright stars and galaxies, although we know that there is some dark matter around in the form of dust and gas, and the best modern ideas of galaxy formation (see Chapter 2) suggest that there may be at least ten times as much dark matter associated with each bright galaxy. Taking the bright matter alone, astronomers can estimate the mass of each galaxy from its brightness, knowing how the brightness of an individual star depends on mass. If they do that, their estimates indicate that the overall density is only about 1 to 2 percent of the amount necessary to close the Universe. With the new understanding of galaxy formation and the indications that all galaxies have superhalos of dark matter, this figure is changed dramatically to 10 to 20 percent of the required density— but that change is not dramatic enough, if you accept the evidence from the deceleration parameter studies that the Universe really is closed. This, in a nutshell, is one of the biggest problems in cos-

---

* Lloyd Motz, *The Universe* (London: Abacus, 1977), p. 57.

mology today, the puzzle of the so-called missing mass. And the puzzle also operates on a smaller scale, since some studies of the movements of galaxies within clusters show that the velocities of individual galaxies, revealed by the Doppler shift, are bigger than the escape velocity from the cluster, unless "extra" matter is hidden somewhere in the cluster.

The consensus among astronomers today is that the Universe is open. Professor Joseph Silk, of the University of California at Berkeley, has summed up the consensus view neatly: "The balance of evidence does point to an open model of the universe, [although] there are weak points in every argument."* But whether or not the Universe is open, the issue itself is far from closed and the arguments continue to rage. Some of the latest observations indeed suggest that all of our ideas about the details of the expansion of the Universe may be in need of revision, since there has been a flood of evidence over the past couple of years suggesting that the movement of our Galaxy deviates a great deal from its simple expanding Universe relationship with other galaxies, because of a strong "local" influence tugging us in the direction of the Virgo cluster, some 30 million light-years away. This local influence implies that a correction must be applied to all our observations of the expanding Universe, and there is at least a hint here that we do not yet have a clear picture of the behavior of the Universe on a scale big enough to allow confident predictions of whether it will expand forever or one day recollapse.

If the Universe is open, and will expand forever, then that is very much the end of the story. Individual stars in individual galaxies will run through their life cycles and die, so that the picture is one of dying, fading galaxies spreading ever further apart in space as the Universe expands. The present remarkable distinction between the hot, bright stars and the cold, dark space between would eventually be lost, and the entire Universe would reach thermodynamic equilibrium at a state of very low density, a dark and almost empty Universe containing only the cold, dark remains of stars, planets, and black holes.

If, however, the Universe is closed, then the story has hardly

---

* See Joseph Silk. *The Big Bang* (San Francisco: Freeman, 1980), p. 309.

begun. With just enough matter to halt the expansion and start a recollapse, our Universe would not turn around until at least 30 thousand million years from now, and would take as long to collapse back to its present-day density. By about 90,000 million years from the present, everything in the Universe, matter and energy, will be crushed back into a singularity, a new fireball. And what happens then? Quite literally, the laws of physics as we know them do not work under such extreme conditions, just as they "only" work in describing the Big Bang from a fraction of a second after the outburst began. Perhaps everything would simply disappear. Or perhaps, the most intriguing speculation in all of cosmology, there would be a "bounce" at the singularity, with the collapse being turned around to initiate a new Big Bang. A new expanding universe would be reborn, phoenixlike, from the ashes of the old, to begin a new cycle of expansion and collapse, with the formation of new stars and galaxies—and planets, some of them bearing life. In such a situation it may be that there has been an infinite number of previous cycles and there will be an infinite number of future cycles, so that we are here as observers of just one link in a continuous chain of creation and destruction. To anyone uncomfortable about the concept of a singular creation—a unique Big Bang—this cyclic universe idea may be comforting. But this still isn't the end of the story.

How much like our own Universe would the other links in the chain be? At first sight it might seem that although they would contain different stars and different galaxies, they would otherwise be very much like our Universe, with the same *kinds* of stars arranged in the same *kinds* of galaxies and just the same opportunities for life as here in our cycle. But if the laws of physics break down at the singularity, can we be sure that the *same* laws of physics will be rebuilt in the next Big Bang? John Wheeler, of Princeton University, argues that such a coincidence would be extremely unlikely. Using a concept known as "superspace" to describe the behavior of the Universe in mathematical terms, making allowance for quantum mechanical effects, Wheeler argues that quantum fluctuations in the superdense collapsed phase of the Universe both cause the bounce, triggering a new wave of expansion, and reset the constants of physics on which the detailed behavior of the new expanding universe will depend.* This would explain why the constants of the laws of nature in our Universe have such curious values—they are seen as the product of quantum fluctuations during the most recent collapse, in the cycle preceding our Big Bang. But it also suggests that other cycles may be very different from our own, and in particular that life may be an unusual feature of the universal cycles.

We are back, indeed, to another form of the anthropic principle. In most cycles of the Universe, conditions would be quite unsuitable for life—the universe might, for example, recollapse into a singularity before life could form, or stars might not process primordial hydrogen into heavy elements and scatter them in the right way for life to make use of them. Hardly any cycles of such an oscillating universe would allow life to develop, and our cycle is seen as one of these very rare freaks. Once again it seems that the Universe we see about us *must* be a very unusual variety, or else we wouldn't be here to notice that it was odd!

So whether the argument is put in this form, with a succession of universes following one another in time, or whether it is possible, as some people argue, that the different probability universes run in some sense alongside each other in time, we always come back to this striking conclusion, that life like us can only exist in a peculiar kind of universe. It was Einstein who remarked that "the most incomprehensible thing about the Universe is that it is comprehensible"; the anthropic principle, in one form or the other, enables us perhaps to understand this comprehensibility—life like us, it seems, can only exist in a Universe which we can understand.

---

* For a technical description of superspace, the definitive source is *Gravitation*, by Charles Misner, Kip Thorne, and John Wheeler (see Bibliography). But it *is* technical!

APPENDIX A
THE AGE OF
THE UNIVERSE

The key to everything else in this book is the belief that we understand how the Universe began, and that we know with reasonable accuracy how long ago this beginning took place. So I include here a report of the very latest view, at the time of writing, on this crucial issue. The material first appeared in a slightly different form in New Scientist, *13 March 1980. It is based on a discussion I had with Dr. David Hanes, then at the Institute of Astronomy, University of Cambridge.*

Professional astronomers were excited, and the general public intrigued and perhaps mystified, by the announcement late in 1979 that a remeasurement of the age of the Universe indicated it to be only half as old as had lately been believed (*New Scientist*, 22 November 1979, p. 587). Three young American astronomers—Marc Aaronson of Steward Observatory, John Huchra of Harvard University, and Jeremy Mould of Kitt Peak National Observatory—had redetermined by a new technique the distances to several relatively nearby galaxies. They suggested as a result that the best estimate of the age of the Universe should lie close to 10,000 million years, rather than the 20,000 million years which has been increasingly accepted by the astronomical community over the past twenty years or so. To find out how simple measurements of the *distances* to galaxies are able to tell us about the age of the Universe and why there should be disagreement over a factor of 2 in the results obtained by different techniques, I spoke to David Hanes of the Institute of Astronomy in Cambridge. He too has been looking at the problem of measuring the distances to galaxies and the implications for our understanding of the Universe at large.

Hanes stressed that animated and often heated discussions concerning just this range of age estimates, 10 to 20 thousand million years, have been going on for the past two decades, and that it was only in 1952 that astronomers settled even on this "ball park" of estimates. It was then that astronomers realized their established scale of cosmic distances was wrong, not by a factor of 2, but by a factor of 4. Virtually overnight the accepted "age of the Universe" at that time was quadrupled—a discovery which captured public attention to an unprecedented degree. It puts the current bout of excitement in perspective as an indication that perhaps astronomers are steadily getting better estimates of when the Big Bang really happened.

A factor of 2 is, in any event, nothing too extravagant when we are dealing in ages of thousands of millions of years. The difference between 10 and 20 thousand million is indeed a factor of 2; but the difference between 1 year and 10 thousand million years is a factor of 10 thousand million! The "error" present in the age estimates is much less significant than the dramatic fact that astronomical observations are able to provide a clear indication that there was a definite beginning to the Universe, however many thousand million years ago it took place.

The history of the concept of the expanding Big Bang Universe is remarkably short. As Hanes stresses, the most meaningful "age of the Universe" we have is the so-called expansion age, and the discovery of the expansion of the Universe dates back only to 1912, when Vesto Slipher, working at the Lowell Observatory, first established that many nonstellar objects—nebulae—are receding from us at velocities of hundreds or thousands of kilometers per second. These are much greater than the measured Doppler velocities of stars in our Milky Way. The recession velocities were measured from the Doppler red shifts of features in the spectra of these nebulae, but the precise nature of the nebulae remained a mystery until 1924, when Edwin Hubble, using the then-new 100-inch telescope at Mount Wilson, detected individual Cepheid variable stars in some of them. He was able to prove beyond doubt that they lay beyond our Milky Way system and formed galaxies in their own right.

Hubble and Milton Humason were able to show over the next few years that the large recession velocities of the external galaxies are correlated with their apparent brightnesses—the fainter galaxies (which presumably look fainter only because they are further away) are receding more rapidly than brighter (nearby) galaxies. This correlation was confirmed and refined by measurements of Cepheid variables in the closer external galaxies—the Cepheids pulsate with a period related to their bightness, giving "standard candles" for making distance measurements over the range that individual Cepheid stars can be identified. The improving measurements soon showed that the relationship between velocity and distance is linear, and it is usually expressed algebraically in the form $V = H_0 d$— the recession velocity V of a galaxy is proportional to its distance d. The constant of proportionality, H_0, is now known as the Hubble constant; Hubble himself used a k in the equation. This discovery, combined with the realization that Einstein's equations could be interpreted as a description of an expanding universe, established cosmology as a science and made it possible eventually to describe how the Universe has evolved from an initial outburst (the Big Bang) to its present state (see *New Scientist*, vol. 83, p. 506).

Leaving aside the cosmological subtleties and looking just at what these discoveries tell us about the age of the Universe, it is important first to ignore the small-scale "peculiar motions" of individual galaxies which slightly disturb the pattern of universal expansion. The nearby Andromeda galaxy (M31), for example, is actually moving toward us. Secondly, Hubble's constant is not, strictly speaking, a constant at all, because the gravitational attraction of all the matter in the Universe must act to slow down the expansion of the galaxies away from one another. Hubble's "constant" should decrease with time as the expansion velocities of all the galaxies decrease with time, and this is why it is given the subscript zero, to imply the *present* value of what some scrupulously correct astronomers prefer to call the Hubble "ratio" or Hubble "parameter." But detectable changes in H_0 occur only over times vastly greater than any human timescale. The parameter, H_0, is expressed in units of km/s/Mpc, so that a galaxy at a distance of 1 Mpc (3.26 million light-years, one megaparsec) is receding at a velocity of H_0 km/s; a galaxy at a distance of 2 Mpc has twice that recession velocity; and so on.

So we live in an expanding Universe. Using the helpful (but, he stresses, not strictly accurate) analogy of an exploding bomb. Hanes likens the outward rush of the galaxies to the outward rush of fragments from the bomb. The fastest moving fragments travel furthest in any given time, so that at some instant after the explosion the observed velocities of the fragments will be proportional to their distances from the site of the explosion, ignoring external factors such as air resistance. Such a "Hubble law" holds regardless of which fragment—which galaxy—is chosen as the reference point from which distances and velocities are measured, as Figure 1 shows.

From this analogy it is easy to see a way of determining the age of the Universe. Simply divide the distance between any two galaxies—fragments—by their relative recession velocity, which gives the time since they were "touching," interpreted as the time since the Big Bang. The real Universe is decelerating, of course, so the relative recession velocity must have been bigger in the past, and this simple calculation must give an overestimate of the age of the Universe, the time since the Big Bang explosion. But a consistent overestimate would be enough to satisfy most cosmologists today. In terms of the Hubble constant, H_0, this estimate of the age of the Universe, dubbed the "expansion age" or the "Hubble time," is simply given by $t = d/V = 1/H_0$.

Although Hubble was well aware of these implications, and had a reliable tool for measuring velocities of galaxies in the form of the red shift, it was much harder to measure distances to other galaxies accurately. In addition, because the nearest galaxies have

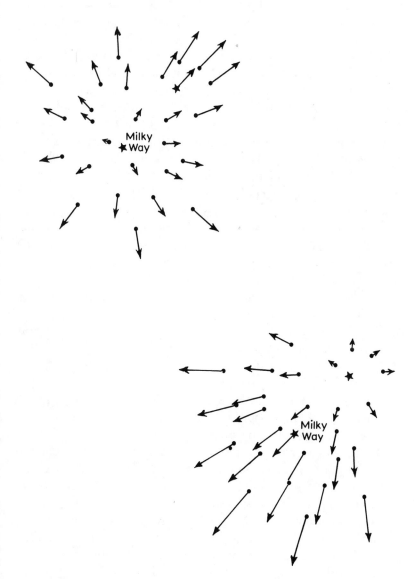

Figure 1 The expansion of the Universe seen from different galaxies always produces a Hubble velocity/distance relationship. (Based on Figure 22 in *Modern Cosmology* by D. W. Sciama)

peculiar velocities which do not fit into the overall expansion pattern, estimates of the expansion age depend on measuring, as accurately as possible, the distances to galaxies at least tens of millions of light-years away. For half a century, the distance estimates have been the weak link in the chain, and it is changes in the distance scale which have led to sometimes dramatic changes in our best estimates of the age of the Universe. Hubble and Humason, pioneers in this work, came up with an estimate of 2,000 million years for the expansion age, embarrassingly short as geological studies already indicated at that time a much older age for the Earth, and nobody was happy with the idea that the Universe might be younger than our planet! The dilemma was resolved in 1952, when it was discovered, with the aid of ever-improving observations, that Hubble had in some cases mistaken one kind of variable star for another and that in some remote galaxies superbright clouds of ionized gas had been misidentified as stars. The result was the fourfold increase in estimates of the extragalactic distance scale which pushed back the expansion age to 8,000 million years, comfortably longer than the age of the Solar System (estimated at around 4,500 million years from various pieces of evidence). Since then, further improvements of the cosmic distance scale have tended to push estimates of the expansion age up toward 20,000 million years—a trend now reversed by the work of Aaronson, Huchra, and Mould, and the completely independent studies made by Hanes, using a relatively unsung, but equally impressive, new technique.

Hanes has tackled the problem by studying the brightness of globular clusters of stars in distant galaxies. These clusters are potentially excellent cosmic distance indicators, not least because they contain very many stars and are therefore very bright, visible over great distances across the Universe. As the name implies, globular clusters are spherical; they may contain up to a million stars in one cluster, with the average brightness of each star comparable with the brightness of our Sun. More than a hundred such clusters are visible around our own Galaxy, and they can be identified in their thousands around galaxies in the Virgo cluster, tens of millions of light-years away. If only all globular clusters were exactly the same brightness, intergalactic distances could be determined simply by measuring their apparent brightnesses, which would be dimmed simply in line with the inverse square law, which says that a "standard candle" at a distance $2x$ from us will appear one-quarter as bright as an identical standard candle at a distance of x. Unfortunately, globular clusters are not all identical. Within our own Galaxy, the brightest globular cluster is more than 100 times as luminous as the faintest one known, and a simpleminded comparison of two globular clusters chosen at random in different galaxies could give hopelessly misleading indications of the relative distances of the two galaxies from us.

But Hanes has found, from a painstaking study of the globular clusters associated with twenty galaxies in the Virgo cluster, that the relative numbers of bright and faint globular clusters in each galaxy follow the same pattern. When the numbers of clusters at each brightness are plotted graphically, as a "luminosity function," essentially identical curves are obtained for galaxies which are known, from their physical association with one another in a group, to be at roughly the same distance from us. When this discovery is applied to studies of the luminosity function of a randomly chosen galaxy's family of globular clusters, it gives an unambiguous indication of distance.

First, Hanes takes the luminosity function for the globulars in our own Galaxy; then he takes the luminosity function for the globulars in the galaxy being studied, and the two plots are compared. Because of the dimming of the light from the distant galaxy by the inverse square law, the two luminosity functions do not look identical in their "raw" state. But by modifying the function which corresponds to the globulars in our Galaxy by an appropriate inverse square diminution, the two curves can be made to coincide exactly. The unique distance that corresponds to the luminosity adjustment required gives the distance to the other galaxy (Figure 4). Astronomers measure the brightness of celestial objects in "magnitudes," so that a distance determined in this way is given in magnitudes as a "distance modulus." Using the Virgo cluster data, Hanes comes up with a distance modulus of 30.7 mag, corresponding to an expansion age of the Universe of 12.5 thousand million years, and a Hubble constant of 80 km/s/Mpc (10 thousand million years corresponds to $H_0 = 100$ km/s/Mpc; 20

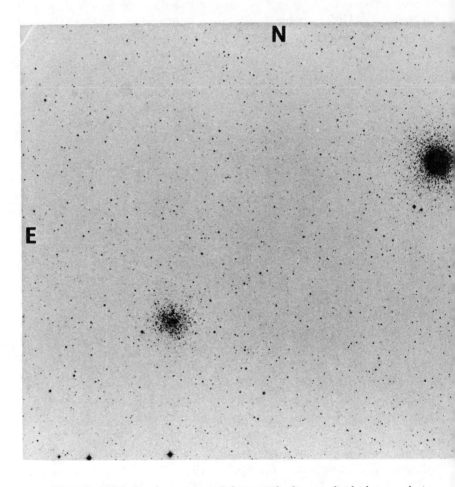

Figure 2 A globular cluster in our Galaxy, resolved into individual stars; galaxies in the Virgo Cluster, are surrounded by thousands of globular clusters that cannot be resolved into separate stars.

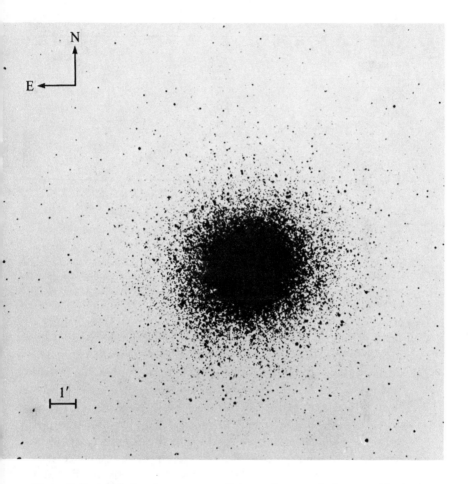

Figure 3 Two globular clusters in our Galaxy. These are both at roughly the same distance from us, so the difference in brightness apparent in the photograph is real.

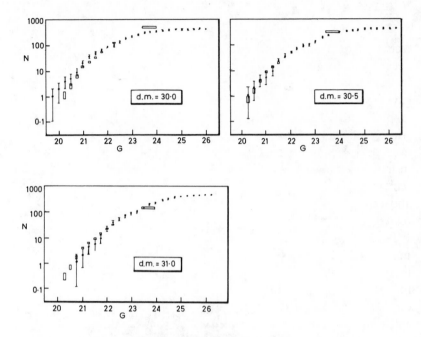

Figure 4 Comparisons of the luminosity function for globular clusters in the five Virgo galaxies (boxes) and in our own Galaxy (dots) with corrections appropriate for shifting our Galaxy's clusters to different distance moduli. The sensitivity of this technique as a distance indicator is clearly indicated; in fact, the best fit of all is obtained for a distance modulus of 30.7 magnitudes.

thousand million years to $H_0 = 50$ km/s/Mpc). These results were published in 1979 in the *Monthly Notices of the Royal Astronomical Society* (vol. 188, p. 901); other techniques, including a mass of evidence collected and analyzed by Allan Sandage and Gustav Tammann at the Hale Observatories, still point to an age of around 20,000 million years, while a minority of other tests give results closer to Hanes's estimate. Although the larger expansion age is the value most widely quoted, there is certainly no consensus among astronomers today; Hanes merely says that "each method has its own merits and problems, and its own adherents and detractors." However, to an outsider looking in on the debate, one thing is clear: whereas the "standard" methods used by Sandage, Tammann, and the others depend on putting several (or many) pieces of evidence together, the globular cluster method uses just one type of observation of one class of astronomical objects. There is intrinsically less room for error in such a method because the longer a chain of reasoning is the greater the probability that it contains a weak link. Aaronson, Huchra, and Mould have also used a one-step technique, but one which enables them to probe further out into the Universe, beyond the distance at which individual globular clusters can be picked out (although even their technique does not cover a truly large volume of space in terms of the whole of the visible Universe; distances to the most remote galaxies can be determined *only* by measuring red shifts and using the best estimate of H_0 in the Hubble velocity/distance relation).

Basically, their technique depends on a very simple fact: more massive galaxies contain more stars and are therefore brighter than their less massive brethren. If the *mass* of a galaxy can be determined, and we know the average brightness of a star, then we can estimate its absolute luminosity and determine its distance from its apparent luminosity and the familiar inverse square law. The mass is estimated by measuring the rate at which the chosen galaxy is rotating, because more massive systems must rotate more rapidly to prevent them from collapsing under their own gravitational pull. For flattened galaxies seen edge on, it is a straightforward matter to measure rotation because the Doppler shifts in spectra taken at the two extremities of the galaxy give a direct indication of how fast one side is moving toward us and the other is moving away (assuming, of course, that the overall red shift caused by the expansion of the Universe is subtracted out). As galaxies are rich in neutral hydrogen gas, Aaronson, Huchra, and Mould actually use the 21-cm radiation of this gas to measure the rotation rates, rather than using optical spectra, but the principle is the same.

So far so good—but it is still necessary to measure the luminosities, and this causes problems. Brent Tully and Richard Fisher, who pioneered the technique, originally measured the luminosities in blue light, but these short wavelengths are easily scattered by dust grains (exactly as blue light is scattered by dust in the atmosphere, leaving a red sunset), and as dust concentrates in the plane of a galaxy, this scattering is particularly important for light coming from galaxies viewed edge on. If face-on galaxies are studied to reduce this difficulty, it is much harder to measure the rotation rates—direct monitoring of the rotation would need observations over thousands of years, because a galaxy like our own takes hundreds of millions of years to turn once. So first attempts to use the Tully-Fisher technique came up with very different results, depending on just what assumptions were made about absorption and how big a correction was applied for inclination effects.

Aaronson, Huchra, and Mould (AHM) have attempted to overcome these difficulties by measuring brightnesses at infrared wavelengths, where the light penetrates dust more easily. This has the bonus that most of the mass of a galaxy is in the form of cool, red, low-luminosity stars, so that infrared observations are close to the wave band where the dominant mass is most visible. (All this discussion applies to the visible, bright galaxies, of course; there is a growing weight of evidence that much more mass may lie outside the bright region as dark matter—*New Scientist*, 8 November 1979, Vol. 85, p. 436). Calibrating their measurements by studying first a few nearby galaxies whose distances are determined very accurately from measurements of Cepheid variables, AHM then turned their attention outward, coming up with the surprises that made headlines late in 1979.

Hanes at least wasn't surprised by the results of the AHM study of the Virgo cluster—a relatively near neighbor on a truly universal scale—as the measurements indicated a distance to that cluster only 14 percent greater than the distance determined from the globular cluster method, agreement well within the uncertainties of the two techniques. But because the AHM technique uses the light of whole galaxies, compared with Hanes's use of the light from clusters within galaxies, it can probe further out still, where it reveals that the Hubble constant determined from the Virgo studies is smaller than the Hubble constant measured for more distant galaxies. It is this discovery, rather than the low expansion age implied overall, that has caused most interest "in the trade," among astronomers. Several other measurements have indicated that our own Galaxy and its immediate neighbors may be falling toward the Virgo cluster at several hundred kilometers per second (see *New Scientist*, 31 January 1980, p. 317), and it looks increasingly as if our Galaxy has such a large "peculiar motion" that accurate measurements of the age of the Universe and conclusions about its evolution will require long-scale mapping over much greater distances than those used so far.

The expansion age derived by AHM is just over 10,000 million years, which is on the edge of being uncomfortably short. Our best theoretical models of stellar evolution suggest that the stars in globular clusters, the oldest known stars, are just about this old and perhaps even older. Hanes suggests that it is possible that a basic rethink of stellar evolution may yet be required if more evidence of such a short expansion age comes in. However, that time has not yet come, and until there is a much narrower spread between the estimates derived by different techniques (the Sandage-Tammann estimate of 20,000 million years still has its adherents) none can be taken as gospel.

A completely independent, and in principle unambiguous, measure of the age of the Universe comes from studies of the decay products of radioactive isotopes found in old material, such as meteorites. As yet, however, the uncertainties involved in the practical application of this technique mean that the corresponding best estimate of the age of the Universe is between 13 and 22 thousand million years (*New Scientist*, 7 February 1980, p. 398). This might suggest that the AHM estimate is rather low, but there is too big a spread to resolve the broad conflict between the AHM and Hanes results and those of Sandage, Tammann, and their colleagues.

The simplicity of both AHM and the Hanes study suggests that they should contain few errors, and the agreement between the two simple techniques is reassuring. The most fascinating question thrown up by all this work, though, is: Just how far out into the Universe do we have to go before truly universal effects completely dominate over such "local" behavior as the movement of our Galaxy, at hundreds of kilometers per second, toward a "neighbor" such as the Virgo cluster, about 30 million light-years away?

APPENDIX B
GROWTH AND LIMITS

In the immediate future the greatest problem facing mankind is to remove the inequalities that cause conflict between people. So many people today have been conditioned by the now discredited "limits to growth" argument of a decade ago that I include here another slightly modified article of mine from New Scientist, *printed to coincide with the end of the 1970s, which puts the "limits" debate in perspective.*

Although pedants may argue that 1980 is the last year of the old decade and not the first of a new, the crystal-ball–gazing that generally goes on at this time of year is receiving an extra stimulus from the change of not one but two digits in the calendar. Just the same thing happened ten years ago, when we were in the midst of doomsday forecasts suggesting that civilization as we knew it would be lucky to survive into the 1980s, let alone the new millennium. The rather obvious failure of the forecasts being made at the end of the 1960s suggests that the present crop of prognostications should be taken with at least a pinch of salt. There is a current fashion for stories of a future world filled with microchips, unemployment for the masses, and a life of futuristic ease for the rich. But is this any nearer the truth than the gloom of 1969?

Ten years ago, in his opening address to the conference inaugurating the Second United Nations Development Decade, U Thant, then Secretary-General of the United Nations, made the following statement:

> I do not wish to seem overdramatic, but I can only conclude from the information that is available to me as Secretary-General that the members of the United Nations have perhaps ten years left in which to subordinate their ancient quarrels and launch a global partnership to curb the arms race, to improve the human environment, to defuse the population explosion, and to supply the required momentum to development efforts. If such a global partnership is not forged within the next decade, then I very much fear that the problems I have mentioned will have reached such staggering proportions that they will be beyond our capacity to control.

Well, here we are a decade later, and that doomsday prediction, like so many others, has proved inaccurate. We still stagger from crisis to crisis, and people still predict the collapse of civilization coming in the "next" ten years. At the other extreme, a few super-optimists predict that if we can just hang on for a few more years then technology will usher in an era of plenty on the wings of a great boom of growth—and they too were saying the same thing ten years ago. But despite the apparent wide gap between the prophets of boom and the prophets of gloom, the futurologists have moved closer together over the past ten years, away from their separate entrenched positions at the extremes.

It was the notorious first report to the Club of Rome, *The Limits to Growth*, which brought the futures debate dramatically into the public eye in the early 1970s. But historically Herman Kahn and his Hudson Institute team were the first of the modern generation of futurologists to make an impact on academic and government circles, with ideas diametrically opposed to those of the *Limits* team. Kahn's view of the good times just around the corner appeared in 1967 in a book *The Year 2000* (New York: Macmillan) which emphasized the view that growth would "proceed more or less smoothly through the next thirty years and beyond." The world of the year 2000 portrayed (on the assumption of no major nuclear war intervening) was one in which all countries become richer, although some more rich more quickly than others—the world of 1967 writ large: more hamburgers! more fast cars! more neon lights! more color TV! and so on.

The book seems almost calculated to leave the unsuspecting reader believing that this development of the world along the path of "Western" affluence is inevitable, and some futurologists have suggested that *The Year 2000* was a propaganda exercise, a deliberate attempt to steer the world in this direction by influencing public opinion. There is no doubt that the book and its projections were taken seriously and have influenced policymaking both by governments and big business in the United States. The dangers are obvious: by presenting one view of the world of the future, one "future world" as inevitable, it may *become* inevitable as we begin to act in the way laid down in the blueprint. So perhaps it is just as well that the opposed extremists from the gloomy camp in the futures debate came along to shake up the Kahnian picture.

Even in 1977, however, when intervening events had made *The Year 2000* look pretty shaky as crystal-ball–gazing, Kahn's team came up with a new contribution to mark the U.S. bicentennial, a book called *The Next 200 Years*. But a curious thing has happened. Although this latest book is much less extreme, with detailed comments much closer to what most people would regard as sanity, Kahn has managed to retain his image as Archpriest of Boom! And while a similar shift has taken gloommongers, such as the Club of Rome workers, away from their extreme pessimism into the middle ground, their disciples too have failed to pick up on the change. The result is that the wild-eyed young people on both sides of the debate (and of both sexes) continue to hurl brickbats at one another without appreciating how close their respective gurus are to speaking the same language.

Only an extreme gloommonger, surely, could now take exception to such comments as:

In our view, the application of a modicum of intelligence and good management in dealing with current problems can enable economic growth to continue for a considerable period of time, to the benefit, rather than the detriment, of mankind. We argue that without such growth the disparities among nations so regretted today would probably never be overcome, that "no growth" would consign the poor to indefinite poverty. . . . We do not expect growth to continue indefinitely. . . . Our differences with those who advocate limits to growth deal less with the likelihood of this change than the reasons for it.

The "reasons for it" put forward in *The Next 200 Years* revolve around the suggestion that we are now living through a great transition in the history of mankind, following the industrial revolution "which began in Holland and England about 200 years ago." Not exponential growth, but growth following a flattened S curve through a demographic transition is the picture favored by the Hudson Institute, change which mimics exponential growth at first but switches over into a new plateau of stability once the impact of revolutionary change has spread through the system. Exponential growth, as Kahn's team points out, "does in fact appear to be stopping now, and not for reasons associated with desparate physical limitations to growth" (Figure 1).

The demographic transition is seen occurring in every country where wealth has increased, life expectancy gone up, and quality of life improved. And the transition has occurred more quickly in each country that has followed the path pioneered by Britain and Europe. In Western Europe and the United States the change took 150 years; in the Soviet Union, 40 years; and in Japan, a recent case, just 25 years, from 1935 to 1960.

Despite this more recent contribution to the futures debate, and often even ignoring recent contributions from the "limits" school as well, the great mass of people who have some vague fear of the way things are going base their views on *The Limits to Growth* itself, or on secondhand reports of that book. The great thing about the book (Potomac Associates, 1972) is that it did spark off a wave

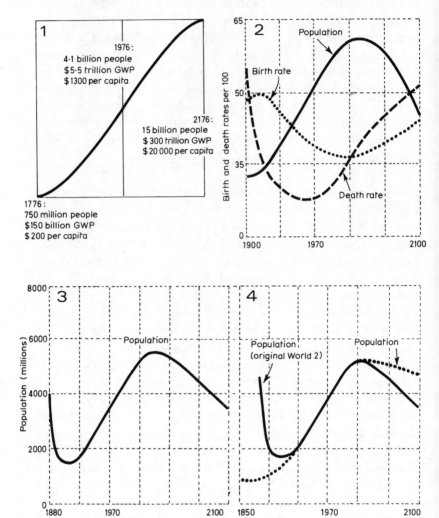

of debate about the future and the nature of the problems facing mankind. But it was neither the first of the major contributions from the modern gloommongers nor the last word even in their eyes; things have moved on since 1972, and the "futures debate" is *not* the same thing as the "limits debate."

If anyone can be credited with parentage of the idea of eco-catastrophe in its modern form, the credit must go to Anne and Paul Ehrlich, who argued in several books published in 1970 and 1971 that the world was even then *already* "over the top" in terms of population and destruction of resources. Far from suggesting more growth, they presented a view that as a matter of urgency the rich must give away their worldly goods to the poor in the hope of ultimate redemption, following the path of "de-development." The basis for such a dramatic turnaround could involve a change in moral attitudes of the kind usually associated with religious conversion—"repent, for the end of the world is at hand." And without such a change they see a future world of famine, plague, and nuclear war.

Rather strangely, alongside this moralistic view, the Ehrlichs also claim that things are already so desperate that the poorest nations are beyond help, and preach the doctrine of "triage"—a nice jargon word disguising the nasty idea of writing off whole nations as beyond realistic help and leaving them to suffer. This really is the counsel of despair; but others on the gloomy side of the fence, if not the Ehrlichs, have more recently been able to discern, if only dimly, the possibility of a reasonably pleasant future world ahead in spite of all our problems.

Before this moderating process occurred, however, the prophets of gloom had a field day, thanks to the arrival on the scene of computer modeling. In the early 1970s people were ready to be told of imminent disaster, and they were also ready to accept as wisdom any calculations carried out by an electronic computer—though why they should be so trusting, when stories of encounters with computerized banking systems, ticket booking offices, and the like have become part of folklore, it is hard to fathom. Perhaps in the early 1980s we will be less trusting. What comes out of the computer depends on what you put in; the catch phrase "garbage in, garbage out" sums it up, though GIGO might as well stand for "gloom in, gloom out." So the *Limits* forecasts were grabbed by many newspapers and nonspecialists as authoritative and blessed by the holy computer, with never a pause to question what was put into the system to get the impressive computer plots out. It is as if you received a bank statement showing an overdraft of $1 million and accepted it without asking how the numbers had been arrived at. And this is *exactly* the wrong approach.

The importance of the work of Jay Forrester, who first developed the computer programs, and of Dennis Meadows's team, which used them to great effect, is that they provide a quick way to see just how the forecasts can be affected by changes in the assumptions that are fed into the model.

Take one curious example. While the forecasts described in *The Limits to Growth* envisage a short-lived boom followed by collapse, the economic growth and population growth during the boom run faster than in the "optimistic" forecasts of Kahn's team! And how can any implications of global inequality—even the desperate implications of the Ehrlichs—be accounted for in a model which uses global averages (for growth, pollution, population, and so on)? The best way to show the true effectiveness of the computer modeling game is to change some of the rules fed in and see what new predictions come out—and that is just what a team from the Science Policy Research Unit (SPRU) at Sussex University did (*Thinking about the Future*, edited by H. S. D. Cole, Chatto and Windus, 1973).

The starting point for this study was a Forrester-Meadows model called "World 2," which produces the curves shown in Figure 2, the dramatic boom and collapse scenario that, for the past seven years, has colored public debate about futures. Clearly you have to start the model running from some date when things such as birthrate, death rate, and economic growth are well known; you then plug in certain assumptions about how technology, pollution, and so on will affect growth, and set the thing off. The *Limits* team started its run at 1900, a nice round number, and published curves matching up well with known history to 1970, then running wild in the near future, although the death rate figures are a bit off in the years just after 1900. One of the delights of models of this kind is that they can be run backward as well as forward, and, indeed, a standard test of their reliability is to run the model backward and forward a few times between the same dates, when it should always produce the same set of curves. So what happens if we run the World 2 model, which gave the impetus to the publicity for the 1970s limits debate, backward from 1900?

The results (Figure 3) are curious, to say the least. By running the model backward from 1900 and then forward again through 1900, up to date and on into the future, it is possible to get a set of curves which agree with all the gloomy "limits" projections. But the world of the twentieth century turns out, according to the model, to be the sequel to a dramatic boom/collapse cycle of the nineteenth century, in which world population peaked before 1880, at a level above the 1970 figure, and then slumped! The catastrophe involved is so dramatic that the model can't be run back any further—as far as World 2 is concerned, the world *began* in 1880 with a huge population—there is a discontinuity which prevents the model being applied to earlier decades.

This is exactly the kind of behavior which hints at a need to adjust the rules by which the model is being run, in order to bring it in line with reality. All that the SPRU team did was to change Forrester's *guess* that changes in average world material standards cause death rates to vary by a factor of 6 to their own *guess* that the effect covers only a factor of 3 in the appropriate equation in the model; and with that one change the model ran smoothly backward to the beginning of the nineteenth century. The new curves (Figure 4) bear a much closer resemblance to real history than the original World 2, and now, although population still declines in the twenty-first century, the *reason* for the decline is a falling birthrate, not an increasing death rate.

Many other guesses put into the model by Forrester and others can be adjusted with equally dramatic effects. Change estimates of agricultural productivity, say, and you change food supply, population, and ultimately everything in the model. It doesn't forecast *the* future, but rather helps to sketch out a range of possible future worlds. Which one we actually reach depends on how we use what we learn from the models to modify our actions, directing attention to the most urgent problems. The model clearly shows, in fact, that we have a choice of future worlds depending on our actions; but by making people think that it is too late to do anything, the modelers can make doom inevitable.

The shift in position of Meadows's team itself seems to reflect a realization of this. In *The Limits to Growth*, readers were told that all of the conclusions were borne out by detailed work which would be made available as a technical report later. In a curious reversal of the usual practice, the popularization (*Limits*) appeared first, with no backing of detailed explanation immediately available. So unambiguous statements in *Limits*, such as "the limits to growth on this planet will be reached in the next hundred years" and "the basic behavior mode of the world system [is the same] even if we assume any number of technological changes in the system" provided the cornerstone of the populist movement based on the *Limits* work. When the technical report eventually became available, with scarcely any publicity and a much smaller readership, it contained much more cautious remarks, such as: "As the runs presented have shown, it is possible to pick a set of parameters which allow material, capital, and population growth to continue through the year 2100."

Try telling a rabid "eco-nut" that this statement came from Meadows's team, after *Limits*, and see who gets called a liar. The disciples still believe, as an act of faith, that "the limits to growth . . . will be reached in the next 100 years."

All too few people appreciate that all the futures debate from about 1965 to 1975 was merely scene setting prior to the main work. The extremists established their positions, shifted slightly like boxers squaring up to one another, and then began sparring. World 2 and other models were recognized by their users (although not, alas, by their disciples) as the first crude attempts to make computer models of where the world is going. Second-generation models are now beginning to take account of regional differences; and where the original Forrester model had 40 equations in it (to describe the whole world!) and the Meadows version 200, today the modelers talk in terms of 100,000 equations describing the world system. Of course, if you put gloom in you still get gloom out, and if you put optimism in you get optimism out. But you can begin to play different games and learn different things about the world system.

The *greatest* threat to global stability is the inequality between haves and have-nots. We *need* growth, at least enough for the have-nots to close the gap, as no one really believes that the rich are going to give up anything, and continuing inequality must lead to both natural disasters (famine, plague) and armed conflict. The problems are to direct growth appropriately, to help the poor out of the poverty trap and through the demographic transition. Only then does any kind of optimistic view of the future become viable.

BIBLIOGRAPHY

Rather than attempting to provide detailed references to every publication relevant to the entire history of the Universe, the origin of life, and the evolution of mankind, I provide here references to recent books and other publications from which the interested reader can work backward to find older relevant material. I also, of course, include the detailed references for the sources of specific scientific ideas discussed in the text, including some that I do not necessarily agree with.

Ardrey, Robert. *African Genesis*. London: Fontana, 1967.
———. *The Territorial Imperative*. London: Fontana, 1969.
———. *The Social Contract*. London: Collins, 1970.
———. *The Hunting Hypothesis*. London: Collins, 1976.
Attenborough, David. *Life on Earth*. London: Collins, 1979.
Black, Rhona. *The Elements of Palaeontology*, rev. ed. Cambridge, Mass.: Cambridge University Press, 1973.
Brandt Commission. *North-South: A Programme for Survival*. New York: Pantheon, 1980.
Bronowski, J. *The Ascent of Man*. London: BBC, 1973.
Carr, B. J., and M. J. Rees. "The Anthropic Principle and the Structure of the Physical World." *Nature*, Vol. 278 (1979) p. 605.
Clark, David. *Superstars*. London: Dent, 1979.

Daly, Martin, and Margo Wilson. *Sex, Evolution and Behaviour*. N. Scituate, Mass.: Duxbury Press, 1978.

Darwin, Charles. *On the Origin of Species*, facsimile ed., Cambridge, Mass.: Harvard University Press, 1964.

Dawkins, Richard. *The Selfish Gene*. London: Oxford University Press, 1976.

Dermott, S. F., ed. *The Origin of the Solar System*. New York: Wiley, 1978.

Desmond, Adrian. *The Hot-Blooded Dinosaurs*. London: Blond & Briggs, 1977.

———. *The Ape's Reflexion*. London: Blond & Briggs, 1979.

Dineley, David. *Earth's Voyage Through Time*. London: Granada, 1973.

Einasto, Jaan, Mikehl Jôeveer, and Enn Saar. "Superclusters and Galaxy Formation." *Nature*, Vol. 283 (1980) p. 42.

Emiliani, Cesare. "Ancient Temperatures." *Scientific American*, Vol. 198, No. 2 (1958) p. 54.

Enever, J. I. "Giant Meteor Impact." *Analog*, Vol. LXXVII, (March 1966) p. 61.

Fall, Michael. "Dissipation, Merging and the Rotation of Galaxies." *Nature*, Vol. 281 (1979) p. 200.

Folsome, Clair Edwin. *The Origin of Life*. San Francisco: Freeman, 1979.

Freeman, C., and M. Jahoda. *World Futures: The Great Debate*. New York: Universe, 1978.

Gribbin, John. *Our Changing Universe*. New York: Dutton, 1976.

———. "The Age of the Universe." *New Scientist* (13 March 1980), p. 844.

———. "Milankovitch Comes in from the Cold." *New Scientist*, Vol. 71 (30 September 1976), p. 688.

———. "What Future for Futures?" *New Scientist*, Vol. 84.

———. *Timewarps*. New York: Delacorte, 1979.

———. *Future Worlds*. London: Abacus, 1979.

Haldane, J. B. S. "Population Genetics." *New Biology*, Vol. 18 (1955) p. 34.

Hays, J. D., John Imbrie, and N. Shackleton. "Variations in the Earth's Orbit: Pacemaker of the Ice Age." *Science*, Vol. 194 (1976) p. 1121.

Hoyle, Fred. *The Black Cloud*. London: Heinemann, 1957.

———, and N. C. Wickramasinghe. *Lifecloud*. London: Dent, 1978.

Imbrie, John, and Katherine Palmer Imbrie. *Ice Ages: Solving the Mystery*. New York: Enslow, 1979.

Irvine, W. M., S. B. Leschine, and F. P. Schloerb. "Thermal History, Chemical Composition and Relationship of Comets to the Origin of Life." *Nature*, Vol. 283 (1980) p. 748.

John, Laurie, ed. *Cosmology Now*. London: BBC, 1973.

King, Marie-Claire, and A. C. Wilson. "Evolution at Two Levels in Humans and Chimpanzees." *Science*, Vol. 188 (1975) p. 107.

Kukla, George. "Missing Link Between Milankovitch and Climate." *Nature*, Vol. 252 (1975) p. 600.

Kummel, Bernhard. *History of the Earth*, 2nd ed. San Francisco: Freeman, 1970.

Leakey, Richard, and Roger Lewin. *Origins*. London: Macdonald & Jane's, 1977.

———, and Roger Lewin. *People of the Lake*. New York: Anchor Press/Doubleday, 1978.

Lorenz, Konrad. *On Aggression*. London: University Paperbacks, 1967.
Lovelock, J. E. *Gaia: A New Look at Life on Earth*. London: Oxford University Press, 1979.
McCrea, W. H. "Solar System as Space Probe." *Observatory*, Vol. 95 (1975) p. 239.
Mann, A. P. C., and D. A. Williams. "A List of Interstellar Molecules." *Nature*, Vol. 283 (1980) p. 721.
Margulis, Lynn. *Origin of Eukaryotic Cells*. New Haven: Yale University Press, 1970.
———. "Symbiosis and Evolution." *Scientific American*, Vol. 225 (1971) p. 48.
———. "Genetic and Evolutionary Consequences of Symbiosis." *Experimental Parasitology*, Vol. 39 (1976) p. 277.
Mason, B. J. "Towards the Understanding and Prediction of Climatic Variations." *Quarterly Journal of the Royal Meteorological Society*, Vol. 102 (1976) p. 473.
Smith, John Maynard. *The Theory of Evolution*. New York: Penguin, 1975.
———. "Evolution and the Theory of Games." *American Scientist*, Vol. 64 (1976) p. 41.
Milankovitch, M. "Mathematische Klimatetore und Astronomische Theorie der Klimaschwontungen." *Handbuch der Klimatologie*, ed. W. Köppen and R. Geiger. Berlin: Borntraeger, 1930.
Misner, Charles, Kip Thorne, and John Wheeler. *Gravitation*. San Francisco: Freeman, 1973.
Morris, Desmond. *The Naked Ape*. London: Triad/Mayflower, 1977.
———. *The Human Zoo*. London: Triad/Panther, 1979.
Motz, Lloyd. *The Universe*. London: Abacus, 1977.
Murdin, Paul, and David Allen. *Catalogue of the Universe*. New York: Crown, 1979.
Napier, Bill, and Victor Clube. "A Theory of Terrestrial Catastrophism." *Nature*, Vol. 282 (1979) p. 455.
Niven, Larry, and Jerry Pournelle. *Lucifer's Hammer*. Mt. Kisco, N.Y.: Futura, 1978.
Ojakangas, Richard, and David Darby. *The Earth Past and Present*. New York: McGraw-Hill, 1976.
Patterson, Colin. *Evolution*. Boston: Routledge & Kegan Paul, 1978.
Reid, G. C., et al. "Influence of Ancient Solar Proton Events on the Evolution of Life." *Nature*, Vol. 259 (1976) p. 177.
Ridley, B. K. *Time, Space and Things*. New York: Penguin, 1976.
Ronan, Colin, ed. *Encyclopedia of Astronomy*. New York: Hamlyn, 1979.
Rowan-Robinson, Michael, J. Negroponte, and J. Silk. "Distortions of the Cosmic Microwave Background Spectrum by Dust." *Nature*, Vol. 281 (1979) p. 635.
Sagan, Carl. *The Dragons of Eden*. New York: Random House, 1977.
Schopf, J. William. "The Evolution of the Earliest Cells." *Scientific American*, Vol. 239, No. 3 (September 1978) p. 84.
Silk, Joseph. *The Big Bang*. San Francisco: Freeman, 1980.
Stares, John. *Molecules of Life*. London: Chapman, 1972.
Suarez, Max, and Isaac Held. "Modelling the Climatic Response to Orbital Parameter Variations." *Nature*, Vol. 263 (1976) p. 461.

Watson, James D. *The Double Helix*. London: Penguin, 1970.

Weinberg, Steven. *The First Three Minutes*. London: Deutsch, 1977.

White, S. D. M., and M. J. Rees. "Core Condensation in Heavy Halos: A Two-stage Theory for Galaxy Formation and Clustering." *Monthly Notices of the Royal Astronomical Society*, Vol. 183 (1978) p. 341.

Whitten, R. C., et al. "Effect of Nearby Supernova Explosions on Atmospheric Ozone." *Nature*, Vol. 263 (1976) p. 398.

Wickramasinghe, Chandra. "Where Life Begins." *New Scientist*, Vol. 74 (21 April 1977) p. 119.

INDEX

Numbers in italics refer to illustrations or captions.

Hubble's law, 18, 22–23, 59–60, 322, 329
Huchra, John, 319–20, 324–31
Hudson Institute, 301, 333–35
human beings: apes and, 277, 285–89; evolutionary status of, 245–46; mammal ancestors of, 271–75; primate ancestors of, 282–91; size of, 309–10. *See also* primates
human life: future of, 301–18; origins of, 263–300
Human Zoo, The (Morris), 299
Humason, Milton, 321–24
hunter-gatherers, 295–98
Hyades cluster, 20–21
hydrogen: in Big Bang, 30, 36, 309; chemical nature of, 156–57, 160, 162, 171; interstellar, 45, 52–54, 77, 176–79
hydrogen alpha light, *103*

I

ice ages, 144–55, 254–55, 273, 291–95, 304–5
identical individuals, genetic odds against, 244–45
infanticide, 296
infinity concept, 12n, 15, 34
insectivores, 274–76
insects, 251–53
intelligence, evolution of, 238, 246, 276, 289, 299–300, 312
intelligent life, conditions for, 92, 97–100, 113, 114–17, 126–28
interglacials, 145, 148, 152–53, 304
interstellar matter, *81*, *82*, *83*, 104–5, 118, 170–87, *179*, *180*, 210–11
ion, definition of, 162–63
iridium, 271
iron, 104–5, 108–13, 212
Irvine, W. M., 185
isothermal fluctuation, 50–52, 54, 57
isotopes, 78–80, 160–64. *See also* radioactive decay
isotropic universe, 10, 26–27, 36

J

Jahoda, Marie, 304n
Jodrell Bank, great disk at, 9
Jupiter, 63, 86, 114–15, 267
Jurassic, 257, 262

K

Kahn, Herman, 301, 333–35, 338
Kalahari Desert, 295–98
"killer instinct," 298
Kukla, George, 154
!Kung, 295–98, 297, 299

L

M

N

V

Valentine, James, 272n
Van Allen radiation belt, 139
Venus, 84, 88, 115–20, 122–23, 126–27, 267
vertebrae, evolution of, 248, 249
vertebrates, 248–53
Virgo cluster, 325, 326, 327, 328, 330–31
volcanoes, 134, 136, 292n
Vredefort Ring, 268–71

W

warm-bloodedness, 259
water, 114–21, 173, 217
wave-particle duality, 157–58
Wegener, Alfred, 129–34
Wheeler, John, 317–18
"white holes," 57, 65
White-Rees model galaxy, 48–54
Whitten, R. C., 266

Wickramasinghe, Chandra, 175–87
Wilkinson, D. T., 31
Wilson, Allan, 283
Wilson, Margot, 240
Wilson, Robert, 25–27, 28–29, 31, 173
"Woodstock" (song), 171n
World Futures (Jahoda and Freeman), 304n
World 2, 336, 338–40
worms, emergence of, 251

Y

year, number of days in, 256
Year 2000, The (Kahn), 334

Z

Zel'dovich, Ya. B., 51
Zeta Orionis, 83